Advanced Technologies for Science and Engineering

(Volume 1)

Intelligent Technologies for Automated Electronic Systems

Edited by

S. Kannadhasan
Department of Electronics and Communication Engineering
Study World College of Engineering
Coimbatore, Tamil Nadu
India

R. Nagarajan
Department of Electrical and Electronics Engineering
Gnanamani College of Technology
Namakkal, Tamil Nadu
India

N. Shanmugasundaram

Department of Electrical and Electronics Engineering
VELS Institute of Science Technology
and Advanced Studies (VISTAS)
Chennai, Tamil Nadu
India

Jyotir Moy Chatterjee

Scientific Research Group of Egypt (SRGE)
Lord Buddha Education Foundation
Asia Pacific University of Technology & Innovation
Kathmandu
Nepal

&

P. Ashok

Symbiosis Institute of Digital
and Telecom Management (SIDTM)
Symbiosis International (Deemed University)
Lavale, Pune, Maharashtra, India

Advanced Technologies for Science and Engineering

(Volume 1)

Intelligent Technologies for Automated Electronic Systems

Editors: S. Kannadhasan, R. Nagarajan, N. Shanmugasundaram, Jyotir Moy Chatterjee & P. Ashok

ISBN (Online): 978-981-5179-51-4

ISBN (Print): 978-981-5179-52-1

ISBN (Paperback): 978-981-5179-53-8

©2024, Bentham Books imprint.

Published by Bentham Science Publishers Pte. Ltd. Singapore. All Rights Reserved.

First published in 2024.

need for a court order if at any point you breach any terms of this License Agreement. In no event will any delay or failure by Bentham Science Publishers in enforcing your compliance with this License Agreement constitute a waiver of any of its rights.

3. You acknowledge that you have read this License Agreement, and agree to be bound by its terms and conditions. To the extent that any other terms and conditions presented on any website of Bentham Science Publishers conflict with, or are inconsistent with, the terms and conditions set out in this License Agreement, you acknowledge that the terms and conditions set out in this License Agreement shall prevail.

Bentham Science Publishers Pte. Ltd.
80 Robinson Road #02-00
Singapore 068898
Singapore
Email: subscriptions@benthamscience.net

**BENTHAM
SCIENCE**

CONTENTS

PREFACE

The major objective of Smart Electronic Systems is to provide a platform where researchers from the fields of hardware and software may work together under one roof to speed up the development of smart electronic systems. Effective and secure data sensing, storage, and processing are essential in today's information age. Modern smart electronic systems meet the criteria of effective sensing, storage, and processing. Effective techniques and software that allow for a quicker analysis and retrieval of necessary data are simultaneously becoming more important. The internet world now includes big data, which comprises large, complex data collections. It is becoming harder to store and analyse the vast amount of structured and unstructured data that has to be collected. With concurrent hardware and software development, the Internet of Things (IoT) and cyber-physical systems (CPS) have been growing to include everyday consumer electronics. The effectiveness and performance of current and next generations of computing and information processing systems depend on advancements in both hardware and software. Some of the focused areas in this field include memristor and memristive systems, advanced 3D IC technologies, design methodologies, and 3D pacing. Others include molecular electronics, biosensors, bio-molecular and biologically inspired computing, nanoelectronics for energy harvesting, spintronics, domain-wall and phase-change memories, and nanoelectronics for energy harvesting. Quantum computing, communication, information processing, circuit, system, and sensor design based on nanoelectronics for critical applications, chip-to-system design, and techniques for electronic design automation (EDA) or computer-aided design (CAD) are the topics addressed in this book. At the same time, programming and efficient calculations for quicker examination and retrieval of crucial data are slowly getting out of style.

Massive amounts of both organised and unstructured data are challenging to store and manage. Post-CMOS technologies, the Internet of Things (IoT), and the cyber-physical system (CPS) have all been advancing simultaneously with synchronous hardware and software developments and have surpassed standard client devices. Future ages of figuring and data processing frameworks, as well as the present generation, will be largely influenced by advancements in both design and programming. In order to exchange information and research discoveries on all facets of electronic systems and digital electronics, it aims to bring together eminent academic scientists, researchers, and research scholars. It also acts as a premier interdisciplinary platform where researchers, practitioners, and educators may discuss and present the most recent findings, challenges, and trends in the fields of analogue and digital electronic systems, as well as their applications and solutions. The most recent developments in the study of solidification as well as the processing and analytical issues the society is facing in the twenty-first century are discussed. We would like to thank everyone who participated on behalf of the editors. First and foremost, we would like to congratulate wish success to the writers, whose wonderful work forms the basis of the book. We would like to use this opportunity to express our gratitude to our family and friends for their support and inspiration while we wrote this book. First and foremost, we give all praise and adoration to our omnipotent Lord for his abundant grace, which made it possible for me to successfully complete this book. For their contributions to this edited book, the authors deserve our sincere thanks. We also want to thank Bentham Science Publishers and its whole staff for making the work possible and giving us the chance to participate in it.

The content of this book is summarized as follows:

1. Chapter 1 presents that motorcycle accidents are a major societal concern in many countries. Due to improper riding behaviours including not wearing a helmet, driving

recklessly, driving while intoxicated, riding while fatigued, etc., the issue persists despite public awareness initiatives. The risk of fatalities and impairments is relatively high as a result of delayed assistance to accident victims. Significant economic and societal repercussions are seen for those involved. As a consequence, several research institutions and significant motorcycle manufacturers have created safety devices to protect riders. A good motorbike safety system is also difficult to implement and expensive. Modern communication technologies are being incorporated into the automobile sector to improve the aid provided to those injured in traffic accidents, speed up the response time of emergency services, and provide them with more information about the incident. If the resources required for each catastrophe could be estimated with more accuracy, the number of deaths may be significantly decreased. According to the proposed plan, every vehicle must have an onboard device that can identify accident situations and report them to an outside control unit, which determines the severity of the issue and gives the necessary resources to address it. The development of a prototype using commercially available equipment shows that this technology may significantly save the amount of time it takes to deploy emergency services after an accident.

2. In Chapter 2, it is discussed that Malaria is a sickness that is brought on by the Plasmodium parasite and spread by the bite of female Anopheles mosquitoes. There are four different types of plasmodium that cause malaria. 1. *Plasmodium Falciparum*, 2. *Vivax Plasmodium*, 3. *Plasmodium Ovale* and 4. *Plasmodium Malariae*. Although there are a number of clinical and laboratory procedures for detecting the presence of malaria, the speed and precision needed to do so are insufficient. In order to assess if malaria is present in human RBCs, we have developed a method in this study that makes use of image processing techniques. The technique also establishes the malarial parasite's stage and intensity.

3. Defocus blur is a very frequent occurrence in photographs taken by optical imaging devices, as discussed in Chapter 3. It could be unwanted, but it might also be a deliberate aesthetic impact, which means it might help or hurt how we see the scenario in the photograph. A partly blurred picture could be segmented into blurred and non-blurred parts for tasks like object detection and image restoration. In this research, we present a robust segmentation technique to distinguish in- and out-of-focus picture areas and a sharpness measure based on the local Gabor maximum edge position octal pattern (LGMEPOP). The suggested sharpness measure makes use of the finding that the majority of local picture patches in fuzzy parts contain a disproportionately lower number of specific local binary patterns than those in crisp regions. In conjunction with picture matting and multiscale fuzzy inference, this measure was used to produce high-quality sharpness maps in this study. Our blur segmentation algorithm and six competing techniques were put to the test using tests on hundreds of partly blurred photos. The findings demonstrate that our method produces segmentation results comparable to those of the state-of-the-art and has a significant speed advantage over the competition.

4. In Chapter 4, analytics is discussed as one of the leading technologies today since data is amassing in all shapes, sizes, and volumes, as well as in a dynamic way. In the age of social media and social networks, predictive analytics is particularly popular as data sources expand from data banks to data rivers. This chapter provides an overview of the fundamentals of analytics as well as some of the current predictive analytical models used in the analytical community, such as multiple regression, logistic regression, and the k closest neighbor model. Having a predictive analytical tool in our toolbox is even more important now that we live in the age of machine learning and artificial intelligence.

5. In Chapter 5, it is discussed how simulation, a particularly versatile and adaptable area of computer science, is used to model and analyse systems for which an analytical solution is either unavailable or challenging to achieve. Because it is simpler than conventional approaches, which are often challenging, simulation is also chosen as a method of system analysis. Because of this, simulation is an area with extensive application and demand, making it fascinating and helpful to include a chapter for studying simulation with a case study of modelling a Queuing system.

6. Chapter 6 claims that virtualization is a cloud computing solution that only requires one CPU to operate. Through virtualization, many computers seem to be operating together. Because it saves time, virtualization focuses mostly on efficiency and performance-related activities. Virtualization of operating systems is the main topic of this essay. It is a customised version of a typical operating system that enables users to run numerous programmes that create a virtual environment to carry out different jobs on the same computer by running other platforms. Based on the amount of work they accomplish and the amount of memory they use, this virtual machine aids in comparing the performance of Type 1 and Type 2 hypervisors.

7. In Chapter 7 it is discussed that cloud computing offers a dynamic paradigm that enables consumers and organisations to acquire a variety of additional services in accordance with their needs. The cloud provides services including data storage, a platform for developing and testing applications, a way to access online services, and more. Maintaining application performance in a cloud environment is a common challenge due to Quality of Service (QoS) and Service Level Agreements (SLA) supplied by service providers to the organisation. Distributing the workload across many servers is the main duty carried out by service providers. By effectively allocating resources inside Virtual Machines, a load-balancing strategy should meet user needs. This study discusses the review of several LB strategies that affect overall performance and the research gap.

8. The existing electronic voting system may be readily hacked, according to Chapter 8. There are several strategies used to prevent malpractice. This study, allows for safe voting and forgoes human interference, resulting in a seamless and secure election process. The voter's face and biometric fingerprints are used in this study's authentication process. With the voter fingerprint information already in this database, the first step in the verification process for an electronic voting system may be simply accomplished. Voter facial recognition using data previously stored in the database is the second phase of verification. The voter may cast his or her ballot and deliver it if two-phase verification is completed. The ballot will then be encrypted. This stops false votes and guarantees accurate voting free of any corruption. We have developed a fingerprint-based voting system that eliminates the need for the voter to provide an ID with all of the required information. A person is permitted to vote if the facts match the registered fingerprints' previously recorded information. If not, a warning notice is sent and the individual is disqualified from casting a vote. The administrator will decode and count the votes during the counting phase of an election.

9. Chapter 9 presents a research that outlines a method for resolving the problem of real-time decision-making in farming that arises from rapid changes in circumstances, such as atmospheric changes, monsoons, insect assaults, etc. The future of agricultural technology is big data collection and analysis in agriculture to maximise operational performance and reduce labour expenses. The Internet of Things will, however, have an impact on a far wider range of industries than just agriculture since there are more IoT-related concepts to understand. The adaptation of IoT's capacity for data collecting on crop attributes and for automated decision-making using data analytics algorithms is the main goal of this project.

10. In Chapter 10, it is discussed that biometrics innovation is still one of the major predictions that combined biosciences and innovation, serving as a tool for criminology and security analysts to develop more accurate, robust, and certain frameworks. Biometrics, when combined with various combination techniques like feature-level, score-level, and choice-level combination procedures, remained one of the most researched technologies. Uni-modular biometrics, such as unique marks, faces, and iris, are followed by multimodal biometrics dependent on. By presenting a similar investigation of frequently used and referred to uni- and multimodal biometrics, such as face, iris, finger vein, face and iris multimodal, face, unique mark, and finger vein multimodal, this paper will attempt to lay the groundwork for analysts interested in biometric frameworks moving forward. This comparative research includes the development of a comparison model based on DWT and IDWT. The method towards combining the modalities also entails applying a single-level, two-dimensional wavelet (DWT) that has been solidified using a Haar wavelet to accomplish the good pre-taking care of to eliminate disruption. Each pixel in the picture is subjected to a different filtering operation in order to determine the Peak-Signal-to-Noise Ratio (PSNR). This PSNR analyzes the mean square error (MSE) to quantify the disruption to hail before the division of the largest data set to the chosen MSE. In the most recent advancement, each pixel's concept is fixed up using the opposing two-dimensional Haar wavelet (IDWT), creating a longer image that is better able to recognise approbation, affirmation, and confirmation of parts. The MATLAB GUI is used to implement the diversions for this enhanced blend investigation, and the obtained outcomes are satisfactory.

11. Chapter 11 presents that performance prediction is the estimate of future performance circumstances based on information from the past and the present. Companies, divisions, systems, procedures, and personnel may all get forecasts. This research focuses on evaluating employee performance in terms of behavior, output, and potential for development. When workers perform effectively for their employers, everyone wins. As a result, forecasting staff performance is crucial for a developing company. To this purpose, we suggest the support vector machine, the decision tree (j48), and the naive Bayes classifier as three machine learning methods. These help forecast an employee's behaviour at work. Based on parameters like accuracy, error loss, and timeliness, the Naive Bayes algorithm outperforms the other two algorithms in terms of their findings.

12. Chapter 12 discusses the idea that the discipline of artificial intelligence (AI), which trains computers to comprehend and analyse pictures using computer vision, remains in its infancy. This is particularly true in the medical industry. Coronary computed tomography angiography, or CCTA, is a well-known non-invasive technique for diagnosing cardiovascular diseases (CD). Pre-processing CT Angiography images is a crucial step in a computer vision-based medical diagnosis. Implementing image enhancement preprocess to reduce noise or blur pixels and weak edges in a picture marks the beginning of the research stages. Using Python and PyCharm(IDE) editor, we can build Edge detection routines, smoothing/filtering functions, and edge sharpening functions as the first step in the pre-processing of CCTA pictures.

13. In Chapter 13, a patient-monitoring smart wheelchair system is developed as an ambulatory assistance for persons with disabilities and for continuous monitoring of the user's vital bodily parameters. Four interfaces—eyeball control, gesture control, joystick control, and voice control—have been created for wheelchair control in order to cater to various limitations. The picture of the eyeball is captured using a camera. In order to make the appropriate decisions based on the location of the eyeball, LabVIEW is employed. The wheelchair movement may also be controlled by the other three modes. Anti-collision mechanisms are implemented using ultrasonic sensors. There is a feature in the wheelchair for

measuring body temperature and heart rate. If any parameter is outside of a safe range, this system will notify the appropriate medical authorities and the wheelchair user's chosen persons. The finished product is an innovative kind of assistive technology that would simplify and reduce stress in the life of its user.

14. Chapter 14 presents that the power grid assaults serve as a reminder that although the smart Internet of Things (IoT) might help us regulate our lightbulbs, it also poses the risk of putting us in the dark if it comes under attack. Many works of literature have recently attempted to address the issues surrounding IoT security, but few of them tackle the serious dangers that the development of quantum computing poses to IoT. Lattice-based encryption, a likely contender for the next post-quantum cryptography standard, benefits from strong security and good efficiency, making it well-suited for IoT applications. In this article, we list the benefits of lattice-based cryptography and the most recent developments in IoT device implementations.

15. Due to newly developed technologies in cars, traffic signal prediction devices are made and discussed in Chapter 15. It teaches users how to maneuver a car safely and effectively. Because of the many ways that drivers are distracted nowadays, the number of accidents is rising alarmingly. This technology lowers the danger of distracted driving, which causes accidents, by helping to recognise and deliver information based on data. The concepts of supervised learning, unsupervised learning, and reinforcement learning are addressed under the classification heading and serve as a major directive as the topic of machine learning is introduced. Machine learning may produce many different types of models, including neural networks, naive Bayes, random forests, support vector machines, clustering, etc. The primary concept of machine learning model training is to divide the data into training, testing, and validation sets. In order to deliver the best machine learning project, this chapter's conclusion accesses machine learning methodologies. The suggested method describes how to recognise traffic signs using a model that combines a classic support vector machine (SVM) with a newer convolutional neural network (CNN). In essence, a CNN model was trained to create this model. Several CNN model topologies, including LeNet, AlexNet, and ResNet-50, may be used in this situation. Later CNN layers' output may be utilised to generate features. The SVM was then used to classify using these characteristics.

16. In Chapter 16, machine learning is used to alter the systems that carry out artificial intelligence-related tasks. (AI). It displays the many ML kinds and applications. It also explains the fundamental ideas behind feature selection techniques, which can be applied to a variety of machine learning (ML) methods, including artificial neural networks (ANN), Naive Bayes classifiers (probabilistic classifiers), support vector machines (SVM), K Nearest Neighbor (KNN), and the greedy algorithm-related decision trees algorithm.

S. Kannadhasan
Department of Electronics and Communication Engineering
Study World College of Engineering
Coimbatore, Tamil Nadu
India

R. Nagarajan
Department of Electrical and Electronics Engineering
Gnanamani College of Technology
Namakkal, Tamil Nadu
India

N. Shanmugasundaram
Department of Electrical and Electronics Engineering
VELS Institute of Science Technology
and Advanced Studies (VISTAS)
Chennai, Tamil Nadu
India

Jyotir Moy Chatterjee
Scientific Research Group of Egypt (SRGE)
Lord Buddha Education Foundation
Asia Pacific University of Technology & Innovation
Kathmandu
Nepal

&

P. Ashok
Symbiosis Institute of Digital
and Telecom Management (SIDTM)
Symbiosis International (Deemed University)
Lavale, Pune, Maharashtra, India

List of Contributors

Annamalai Solayappan	Sri Subramaniya Swamy Government Arts College, Tirutttani, Tamil Nadu, India
A. Wisemin Lins	Vels Institute of Science, Technology and Advanced Studies, Pallavaram, Chennai, Tamil Nadu, India
A. Firos	Department of Computer Science and Engineering, Rajiv Gandhi University, Rono Hills, Doimukh-791112, India
Dhivya R.S.	School of Computing Sciences, VISTAS, Thiruthangal Nadar College, Chennai, India
E.N. Ganesh	Vels Institute of Science, Technology and Advanced Studies, Pallavaram, Chennai, Tamil Nadu, India
G. Rajakumar	Department of Electronics and Communication Engineering, Francis Xavier Engineering College, Tirunelveli, Tamil Nadu, India
J. Mary Ramya Poovizhi	Department of Computer Science, Vels Institute of Science, Technology and Advanced Studies (VISTAS), (Deemed to be University), Chennai, India
J. Jebathagam	Department of Computer Science, Vels Institute of Science Technology and Advanced Studies, (VISTAS), Chennai, India
J. Vijayarangam	Department of Applied Mathematics, Sri Venkateswara College of Engineering, Sriperumbudur, Tamil Nadu, India
J. Jebathangam	Department of Computer Science, Vels Institute of Science Technology and Advanced Studies, VISTAS, Chennai, India Department of Information Technology, Vels Institute of Science, Technology and Advanced Studies (VISTAS), Chennai, India
K. Sharmila	Department of Computer Science, Vels Institute of Science, Technology and Advanced Studies (VISTAS), Chennai, India
M. Vidhya	Department of Computer Science, Vels Institute of Science, Technology and Advanced Studies (VISTAS), (Deemed to be University), Chennai, India
M. Gunasekaran	Department of Computer Science, Government Arts College, Salem-636007, India
M. Nisha	Department of Computer Science, Vels Institute of Science Technology and Advanced Studies, VISTAS, Chennai, India
N. Janaki	Vels Institute of Science, Technology and Advanced Studies, Pallavaram, Chennai, Tamil Nadu, India
P. Bhargavi Devi	Department of Computer Science, Vels Institute of Science, Technology and Advanced Studies (VISTAS), Chennai, India
P. Govindasamy	Vels Institute of Science, Technology and Advanced Studies, Pallavaram, Chennai, Tamil Nadu, India
R. Sebasthi Priya	Department of Mathematics, University College of Engineering, Tiruchirappalli, India
R. Kabilan	Department of Electronics and Communication Engineering, Francis Xavier Engineering College, Tirunelveli, Tamil Nadu, India

R. Ravi	Department of CSE, Francis Xavier Engineering College, Tirunelveli, Tamil Nadu, India Department of Electronics and Communication Engineering, Francis Xavier Engineering College, Tirunelveli, Tamil Nadu, India
R. Devi	Department of Computer Science, Vels Institute of Science, Technology and Advanced Studies (VISTAS), (Deemed to be University), Chennai, India
R. Mallika Pandeeswari	Department of Electronics and Communication Engineering, Francis Xavier Engineering College, Tirunelveli, Tamil Nadu, India
S. Shargunam	Department of Electronics and Communication Engineering, Francis Xavier Engineering College, Tirunelveli, Tamil Nadu, India
Seema Khanum	Department of Computer Science and Engineering, Rajiv Gandhi University, Rono Hills, Doimukh-791112, India
S. Kamalakannan	Dept of Information Technology, Vels Institute of Science, Technology and Advanced Studies, Chennai, India
S.V. Rajiga	Department of Computer Science, Government Arts College, Dharmapuri-636705, India
Sujatha P.	School of Computing Sciences, VISTAS, Thiruthangal Nadar College, Chennai, India
S.K. Piramu Preethika	School of Computing Sciences, VISTAS, Pallavaram, Chennai, Tamil Nadu, India
T. Karthikeyan	Dept of Mathematics, Ramakrishna Mission Vivekananda College, Chennai, India
T.R. Premila	Vels Institute of Science, Technology and Advanced Studies, Pallavaram, Chennai, Tamil Nadu, India
T. Santhi Punitha	School of Computing Sciences, VISTAS, Pallavaram, Chennai, Tamil Nadu, India
U. Lathamaheswari	Department of Computer Science, Vels Institute of Science Technology and Advanced Studies, (VISTAS), Chennai, India
V. Jeevitha	Department of Computer Science, Vels Institute of Science, Technology and Advanced Studies (VISTAS), Chennai, India

Innovative Device for Automatically Notifying and Analyzing the Impact of Automobile Accidents

R. Kabilan[1,*], **R. Ravi**[1,2], **R. Mallika Pandeeswari**[1] and **S. Shargunam**[1]

[1] *Department of Electronics and Communication Engineering, Francis Xavier Engineering College, Tirunelveli, Tamil Nadu, India*

[2] *Department of CSE, Francis Xavier Engineering College, Tirunelveli, Tamil Nadu, India*

Abstract: In many nations, motorcycle accidents are a big public issue. Despite public awareness campaigns, the problem continues to grow as a result of poor riding habits such as riding without a helmet, dangerous driving, drunk driving, riding without enough sleep, and so on. Because of late help to those who have been in accidents, the rate of deaths and disabilities is quite high. People who are implicated suffer significant economic and social consequences as a result of them. As a result, various research organizations and large motorcycle companies have developed safety systems to safeguard riders from harm. Furthermore, a dElectronics and Communication Engineeringnt motorcycle safety system is hard to execute and costly. The integration of modern communication technologies into the automotive industry allows for greater assistance to people wounded in traffic accidents, a reduction in the time it takes emergency services to respond, and an increase in the amount of knowledge they have about the occurrence. The number of fatalities might be greatly reduced if the resources necessary for each disaster could be determined more precisely. The developed scheme calls for every vehicle to be equipped with an on-board unit that detects and reports accident scenarios to an exterior control unit that assesses the depth of the problem and provides the needed resources to aid it. The creation of a prototype based on off-th--shelf equipment indicates that this technology can considerably reduce the time it takes to send emergency services following an accident.

Keywords: Buzzer, GSM, GPS, LCD, Tilt Sensor, Zigbee.

INTRODUCTION

An embedded server is a computer that is designed to do one or a few specific functions, frequently under time limitations. It is frequently incorporated as part system, including a PC, on the other hand, is built to be adaptable and suit a huge

---

* **Corresponding author R. Kabilan:** Department of Electronics and Communication Engineering, Francis Xavier Engineering College, Tirunelveli, Tamil Nadu, India; E-mail: rkabilan13@gmail.com

S. Kannadhasan, R. Nagarajan, N. Shanmugasundaram, Jyotir Moy Chatterjee & P. Ashok (Eds.)

of a larger device that includes hardware and mechanical components. An overall spectrum of end-user requirements. Many modern devices are controlled by embedded systems.

One or more primary processing cores, usually a microcontroller or a digital signal processor, control embedded systems. The most important attribute, however, is that it is dedicated to completing a certain task, which may necessitate the use of extremely powerful processors. Because the embedded system is dedicated to a single purpose, design engineers may optimise it to decrease the product's size and cost while improving its dependability and performance. Certain embedded systems are high-density to take advantage of economies of scale. The overall number of cars on the road has increased dramatically in recent decades, increasing traffic congestion and the incidence of traffic accidents. In the event of a traffic collision, finishing the aid of badly wounded passengers within an hour following the occurrence is critical to limiting the detrimental consequences to the occupants' health [1-5].

In terms of the second goal, the efficacy of assistance to passengers engaged in a traffic accident may be increased if rescue services had access to information about the circumstances in which the accident occurred before arriving at the accident scene. This additional data, collected by sensors within the car, would be used to determine the extent of the occupants' injuries. People who are implicated suffer enormous social and economic consequences as a result of them. As a result, various research organizations and major vehicle manufacturers have developed safety systems to safeguard riders from harm. However, decent vehicle safety technology is difficult to deploy and costly.

GPS, crash sensor, tilt sensor, GSM, Zigbee, buzzer, battery, and LCD are all needed for our project. The accident was discovered using a crash sensor. The accident severity is estimated using the tilt sensor. If an accident is discovered, it uses GPS to coordinate and communicate information to the care center as well as the family through GSM and Zigbee. If an accident happens, the vehicle's motor will come to a complete stop. The battery stores the energy generated [6-10]. The received values can be seen on the LCD, and a bell will alert the people around you.

Using an ad-hoc manner, accident information will be communicated to the other car through a Zigbee device. As a result, avoiding a collision is relatively simple. It is simple to determine the overall intensity of the accident. As a result, various research organizations and major vehicle manufacturers have developed safety systems to safeguard riders from harm [11-14].

LITERATURE SURVEY

Prototyping an Automatic Notification Scheme for Traffic Accidents in Vehicular Networks

The proposed system calls for each vehicle to be provided with an on-board unit that detects and reports accident scenarios to an exterior control unit that assesses the severity of the problem and allocates the required resources to aid it. The creation of a prototype using off-the-shelf components and the integration of modern communication technologies into the automotive industry allows for greater assistance to people wounded in traffic accidents, as well as a reduction in the time it takes emergency services to respond and an increase in the amount of evidence they have about the occurrence. The majority of casualties might be reduced if the available resources necessary for each disaster could be determined more precisely. The proposed system calls for every vehicle to be equipped with an on-board unit that detects and reports accident scenarios to an exterior control unit that assesses the severity of the problem and allocates the required resources to aid it.

Emergency Services in Future Intelligent Transportation Systems Based on Vehicular Communication Networks

We've used a technology to increase the pace of operations and boost productivity throughout the years. We've also seen the convergence of computing and telecommunications. This wonderful combination of two critical disciplines has boosted our capabilities, even more, enabling us to communicate at any time and from any location, greatly boosting our workflow flow and raising our quality of life. The confluence of telecommunications, computers, wireless technologies, and advances in transportation is the next wave of development we expect to witness. It may also make this country less enticing to foreign investors since motorists spend more time waiting on clogged highways, causing more pollution.

National Highway Traffic Safety Administration

The Department of Transportation, the NHTSA, and the people we serve all prioritizes safety. We can't believe that highway-related deaths are inevitable. To that aim, the Department is currently working on a significant cross-modal project to produce a national roadway safety plan that will help to address this critical public health concern. Despite changes in road user demands and expectations, the NHTSA is dedicated to its tradition of safeguarding vulnerable road users. The National Highway Traffic Safety Administration will also continue to execute the Department of Transportation's Motorized Coach Safety Action Plan, which encompasses occupant security, structural rigidity, roll avoidance, and fire safety, including emergency evacuation.

Geometry and Motion-based Positioning Algorithms for Mobile Tracking in NLOS Environments

This study proposes positioning methods enabling cellular network-based vehicle tracking in strong NLOS propagation circumstances. The algorithms' goal is to improve network-based positioning systems' positional accuracy when the GPS receiver is really not performing effectively owing to the complicated propagation environment. A one-step and a two-step technique for predicting position are presented and developed. The NLOS impact is greatly decreased by using constrained optimization to optimize the cost function that takes into consideration the NLOS mistake. The created algorithms are practical, which means they may be used in real-world automotive applications. It was also proved through simulation that once the system of constraint is moving at a speed and the heading estimate errors are rather substantial, the one-step technique and the two-step method achieve fairly similar position accuracy.

PROPOSED SYSTEM

GPS, Crash sensor, GSM, Tilt sensor, Zigbee, buzzer, LCD and solar panel are all needed to make our suggested system work. The accident is discovered using a crash sensor. The accident severity is estimated using the tilt sensor. If an accident is discovered, it uses GPS to coordinate and communicate information to the care centre as well as family through GSM and Zigbee. If an accident happens, the vehicle's motor will come to a complete stop. Fig. (**1**) shows the proposed vehicle section.

Block Diagram for Vehicle Section

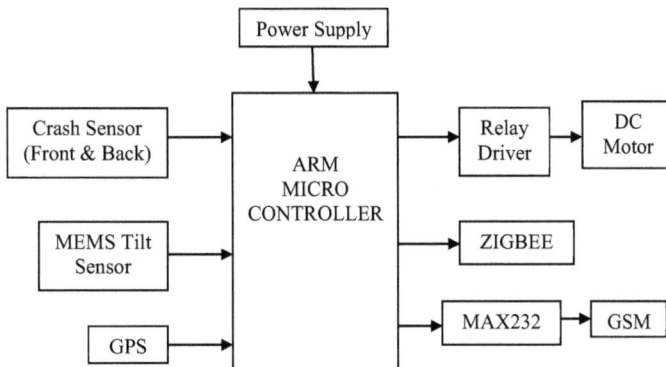

Fig. (1). Vehicle section.

Block Diagram for Other Vehicle Section

Above Fig. (**2**) shows the block diagram of other vehicle section and Fig. (**3**) shows the Circuit diagram of vehicle section.

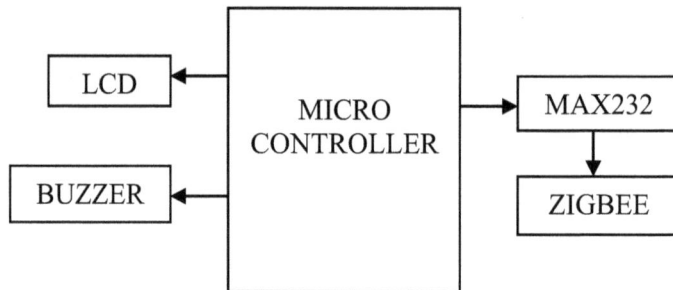

Fig. (2). Other car section.

ARM

ARM is a series of instruction set designs enabling computer processors based on the British firm ARM Holdings' RISC design. This method saves money, heat, and energy. A simplified architecture allows for effective multi-core CPUs and larger core counts at a lower cost, resulting in increased server energy efficiency. Apple, Applied Micro, Atmel, Broadcom, Freescale Semiconductor, Qualcomm, Samsung Electronics, ST Microelectronics, and Texas Instruments are among the companies that use the ARM architecture.

Crash Sensor

The crash sensor could be processed at several places across the vehicle. The amount of velocity they are intended to detect determines their position. They'll tell the SRS control unit if there's been a side collision and if the left or right side airbags should deploy. The driver and passenger airbags would engage if such an accident were to happen from the front.

MEMS Sensor

Modern accelerometers are typically tiny MEMS, and they are the most basic MEMS devices available, comprising nothing but a cantilever beam with just a proof mass. The leftover gas, enclosed in the apparatus, causes damping. As provided, damping affects sensitivity as the Q-factor may not be that low.

DC Motor

The operation of any electric motor is based on basic electromagnetism. When a current-carrying conductor is put in a static electrical field, it creates a magnetic field that produces a force according to the current that flows and the strength of the external magnetic field. A DC motor's internal arrangement is designed to create rotational motion by harnessing the magnetic coupling between current-carrying conductors and an external magnetic field. As you may recall from fundamental electrical theory, the field winding of a shunt motor is coupled into an armature circuit rather than in series with the armature of a series motor. A parallel circuit was widely termed a shunt.

Circuit Diagram for Vehicle Section

Fig. (3). Vehicle section.

Relay

Relays seem to be switches that are powered by electricity. Although other simpler ways are sometimes used, as demonstrated in shown in Fig. **(3)**, an electromagnet is often used in relays to mechanically activate a switching mechanism. The first relays were employed in long-distance telegraph circuits, repeating and re-transmitting the signal from one circuit to another. Relays were widely employed to conduct logical operations in telephone exchanges and primitive computers.

MAX 232

It's an integrated circuit (IC) that translates signals from just an RS-232 serial port into signals that may be used in TTL equivalent digital logic circuits. It was initially introduced in 1987 by Maxim Integrated Products. The MAX232 is a double driver/receiver that transforms the RX, TX, CTS, and RTS signals in most cases.

GSM

A GSM modem is a type of wireless modem that connects to a GSM network. A wireless modem functions similarly to a dial-up modem. The major distinction would be that a dial-up modem transmits and receives data over a fixed telephone line, whereas a wireless modem transmits and receives data using radio waves. The GSM modem's operating principle is based on instructions that always begin with "AT" and end with a "CR" character.

GSM Interfacing with ARM

The U.S. Department of Defense created the GPS, which is a GNSS. It is the world's only fully operational GNSS. It makes use of a network of 24 to 32 low-earth-orbit satellites that send out accurate microwave signals that allow GPS receivers to identify their present location, time, and velocity. NAVSTAR GPS is its official name. The function of a GPS receiver is to find some or all of these satellites, calculate their distances, and use that information to determine their location. It is a simple mathematical idea that underpins this procedure.

ZIGBEE

The ZigBee specification defines a technology that is meant to be simpler and less costly than existing WPANs, including Bluetooth. ZigBee is designed for low-data-rate, long-battery-life, and secure networking radio-frequency (RF) applications. In other words, ZigBee is designed to connect with power line networking, if only for advanced grids and smart appliance applications.

The ZigBee protocols are designed for embedded applications that need low data rates and power consumption. The present goal of ZigBee is to design a general-purpose, low-cost, identity mesh network that can be used for factory automation, embedded sensing, medical data gathering, smoke and intrusion detection, building automation, and home automation, among other applications.

BUZZER

A buzzer is an electrical signaling device that is commonly seen in vehicles, home goods such as microwave ovens, and game shows. Nowadays, a ceramic-based piezoelectric sound wave that produces a high-pitched tone is becoming increasingly prevalent. These were usually connected to "driver" circuits that changed the strength of the signal or pulsed it on and off.

RESULTS AND DISCUSSION

Fig. (**4**) shows that the GPS and GSM module that can be interfaced with the ARM controller.

Fig. (4). GPS and GSM interfacing with ARM.

Fig. (**5**) shows the vehicle section that can be placed on the 4wheel frame which is made for a car.

Fig. (5). The Frame.

Fig. (**6**) shows the LCD that can be interfaced with ARM which is used to display messages.

Fig. (6). LCD Interfacing with ARM.

Fig. (**7**) shows the LCD interfacing with a microcontroller to display the message.

Fig. (7). LCD Interfacing with Microcontroller.

Fig. (**8**) shows that the power supplies that can be interfaced with the GSM module.

Fig. (8). Power supply.

Fig. (**9**) shows the other vehicle section.

Fig. (9). Other vehicle section.

Fig. (**10**) shows the full portion of the vehicle section and other car sections.

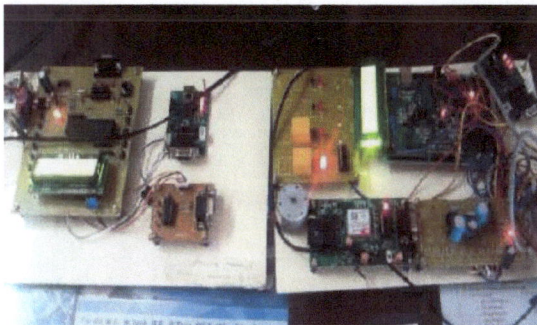

Fig. (10). Full portion kit of vehicle and other car section.

Fig. (**11**) shows the result of the vehicle section that can be displayed on the LCD screen.

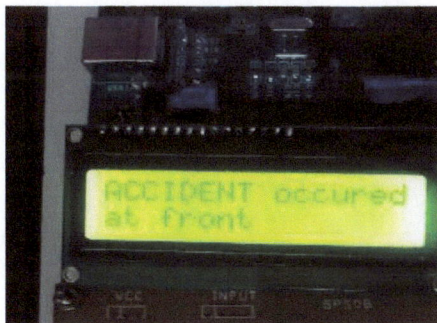

Fig. (11). Result displaying LCD.

Fig. (**12**) shows that the message can be sent to the mobile phone from the vehicle section.

Fig. (12). Mobile phone.

Fig. (**13**) shows that the latitude and longitude values can be displayed on the LCD screen detected by GPS.

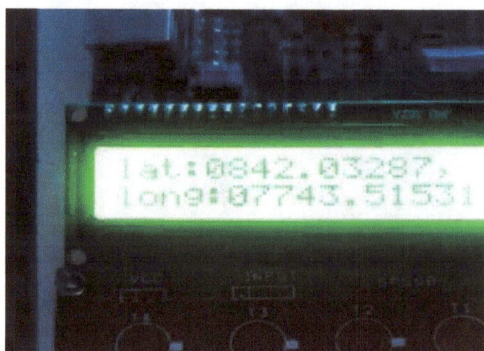

Fig. (13). Result displaying LCD.

Fig. (**14**) shows that the receiving section of the other car section demonstrates how the message is displayed on an LCD screen by the other car's receiving portion.

Fig. (14). Result displaying LCD.

CONCLUSION

In this chapter, we presented the e-NOTIFY system, which allows fast detection Using a combination of V2V and V2I communications, the e-NOTIFY system, which we introduced in this chapter, enables quick detection of traffic accidents, improving the assistance of injured passengers by speeding up the response time of emergency services, and the submission of pertinent information on the accident's conditions. This design substitutes the conventional procedures for accident notification, which rely on witnesses who could provide inaccurate or partial information at the wrong moment. The creation of a low-cost prototype demonstrates that it is possible to integrate this technology into current cars on a large scale. We tested our prototype at the Passive Security Division of Applus+ IDIADA Corporation and demonstrated how it can quickly and accurately send all the specific details of a traffic accident to a Control Alert System. Future work in this area will include installing the OBUs in actual automobiles and deploying the system to test how it responds to rapid acceleration.

ACKNOWLEDGEMENTS

Financial support obtained from the All India Council for Technical Education (AICTE) under Research Promotion Scheme (RPS), Sanction order no: F.No 8.9/RIFD/RPS/Policy- 1/2017-18 coordinated by Anna University Recognized Research Centre, Department of Computer Science and Engineering, Francis Xavier Engineering College, Vannarpettai, Tirunelveli 627003, Tamil Nadu, India.

REFERENCES

[1]   Available from: http://www.dgt.es/portal/es/seguridad vial/estadistica

[2]   G. Fang Wu, F. Jia Liu, and G. Liang Dong, "Analysis of the influencing factors of road environment in road traffic accidents", *2020 4th Annual International Conference on Data Science and Business Analytics (ICDSBA),* pp. 83-85, 2020.
[http://dx.doi.org/10.1109/ICDSBA51020.2020.00028]

[3]   A. Poliaková, "What is the impact of road tax collection on the accident status due to the fault of the road?", *2018 XI International Science-Technical Conference Automotive Safety,* pp. 1-5, 2018.
[http://dx.doi.org/10.1109/AUTOSAFE.2018.8373336]

[4]   C.K. Wong, "Designs for safer signal-controlled intersections by statistical analysis of accident data at accident blacksites", *IEEE Access,* vol. 7, pp. 111302-111314, 2019.
[http://dx.doi.org/10.1109/ACCESS.2019.2928038]

[5]   C. Zhang, Y. Shu, and L. Yan, "A novel identification model for road traffic accident black spots: A case study in ningbo, china", *IEEE Access,* vol. 7, pp. 140197-140205, 2019.
[http://dx.doi.org/10.1109/ACCESS.2019.2942647]

[6]   X. Shen, and S. Wei, "Application of xgboost for hazardous material road transport accident severity analysis", *IEEE Access,* vol. 8, pp. 206806-206819, 2020.
[http://dx.doi.org/10.1109/ACCESS.2020.3037922]

[7]   M. Manzoor, M. Umer, S. Sadiq, A. Ishaq, S. Ullah, H.A. Madni, and C. Bisogni, "RFCNN: Traffic accident severity prediction based on decision level fusion of machine and deep learning model", *IEEE*

Access, vol. 9, pp. 128359-128371, 2021.
[http://dx.doi.org/10.1109/ACCESS.2021.3112546]

[8]   G. Xie, A. Shangguan, R. Fei, X. Hei, W. Ji, and F. Qian, "Unmanned system safety decision-making support: Analysis and assessment of road traffic accidents", *in IEEE/ASME Transactions on Mechatronics,* vol. 26, no. 2, pp. 633-644, 2021.
[http://dx.doi.org/10.1109/TMECH.2020.3043471]

[9]   Available from: http://www.idiada.es

[10]   F. Martinez, C.-K. Toh, J.-C. Cano, C. Calafate, and P. Manzoni, "Emer-.gency services in future intelligent transportation systems based on vehicular communication networks", *Intelligent Transportation Systems Magazine, IEEE,,* vol. 2, no. 2, pp. 6-20, 2010.

[11]   Available from: http://code.google.com/apis/maps

[12]   Available from: http://www.euroncap.com/testprocedures.aspx

[13]   Available from: http://www.rcar.org

[14]   Q. Cai, "Cause analysis of traffic accidents on urban roads based on an improved association rule mining algorithm", *IEEE Access,* vol. 8, pp. 75607-75615, 2020.
[http://dx.doi.org/10.1109/ACCESS.2020.2988288]

CHAPTER 2

Detection of Malarial Using Systematized Image Processing

S. Shargunam[1,*] and G. Rajakumar[1]

[1] *Department of Electronics and Communication Engineering, Francis Xavier Engineering College, Tirunelveli, Tamil Nadu, India*

Abstract: The disease, Malaria, is caused by Plasmodium Parasite, and is transmitted *via* female Anopheles mosquito bite. There are 4 variants of plasmodium which cause malaria, they are, 1) *Plasmodium falciparum*, 2) *Plasmodium vivax*, 3) *Plasmodium ovale*, and 4) *Plasmodium malariae*. Though there are several clinical and laboratory techniques for finding the presence of malaria, the accuracy and the time required to determine the presence of the parasite are inadequate. Therefore, in this work, we have come up with a system that uses image-processing techniques to determine the presence of malaria in Human RBCs. In addition, the system determines the severity and stage of the malarial parasite.

Malaria is brought on by the Plasmodium parasite and spread *via* female Anopheles mosquito bites. *Plasmodium falciparum*, *Plasmodium vivax*, *Plasmodium ovale*, and *Plasmodium malariae* are the four plasmodium species that cause malaria. Although there are a number of clinical and laboratory methods for detecting the presence of malaria, the speed and precision needed to do so are insufficient. As a result, in this study, we have developed a system that employs image-processing methods to ascertain if there is malaria present in human RBCs. The technique also establishes the malarial parasite's stage and intensity.

Keywords: Greyscale, Image processing, Malaria, Thresholding.

INTRODUCTION

The bite of an infected female anopheles mosquito transmits the terrible sickness known as malaria. Over 1 million people die and an estimated 300–400 million people are afflicted each year. Besides, 40% of the population lives in malaria-risk areas. Bashir *et al.* [1] introduced a model for diagnosing malaria using images of stained thin blood smears. The technique makes use of the intensity characteristics of erythrocytes and Plasmodium parasites. Photos of both infected

* **Corresponding author S. Shargunam:** Department of Electronics and Communication Engineering, Francis Xavier Engineering College, Tirunelveli, Tamil Nadu, India; E-mail: shargunamguna@gmail.com

S. Kannadhasan, R. Nagarajan, N. Shanmugasundaram, Jyotir Moy Chatterjee & P. Ashok (Eds.)

and uninfected erythrocytes were obtained, pre-processed, and significant features were extracted from them; ultimately, a diagnosis was performed using the features derived from the images. An artificial neural network (ANN) classifier was used to test the performance of a set of intensity-based characteristics on samples of red blood cells from the newly constructed database. These traits could be successfully applied for malaria detection, according to the results.

Mehrjou *et al.* [2] created a motorised, sophisticated microscope with an autonomous system that can snap high-speed images of blood smears. The main task of our work in this study is image processing, which is started once there are enough samples for microscopy. Saputra *et al.* [3] provide a thorough analysis of the conventional expert and computer-aided identification methods used to diagnose malaria in thin blood smears. As a pre-processing technique for segmentation, Reni *et al.* [4] introduced a new approach for morphological filtering of blood pictures. In blood pictures, conventional morphological closure removes the undesirable elements but also results in the loss of important information. The suggested morphological filtering keeps the important foreground component information while eliminating noise and artefacts. The pre-processing of additional pathological images using this technique, such as tissue analysis and cell differential analysis, could be adjusted. Das *et al.* [5] presented morphological and textural data-based computer-assisted malaria infection prediction, specifically for *Plasmodium vivax*. Here, erythrocytes have been separated from peripheral blood smear light microscopic images employing marker controlled watershed and pre-processing. In order to diagnose malaria using vision alone, Vink *et al.* [6] suggested a simple, quantitative cartridge-scanner system that focuses on low parasite densities. To generate a thin blood film, we employed specialised finger-prick cartridges filled with acridine orange, and a specialised scanner to picture the cartridge. We have created a *Plasmodium falciparum* detector using supervised learning. Dallet *et al.* [7] present a platform for an Android mobile application that enables quick and accurate diagnosis of malaria from thin blood film pictures stained with Giemsa. The innovative Annular Ring Ratio Method, which has previously been implemented, tested, and validated in MATLAB, serves as the foundation for the application. The technique finds the parasites in the infected RBCs as well as other blood components including White Blood Cells (WBCs) and Red Blood Cells (RBCs). The programme also recognises the various stages of parasite life cycles and computes parasitemia, a measurement of the degree of infection. By putting up a brand-new technique for identifying blood components termed the Annular Ring Ratio transform, he made a contribution to the field of mathematical morphology. Additionally, it has suggested an automated algorithm for separating white blood cells from red blood cells that, when combined with the ARR transform approach, has numerous uses for blood-related analyses including microscopic examination in addition to helping to diagnose malaria [8]. Makkapati *et al.* [9] provided a method for detecting parasites in microscope images of blood smears based on image processing, as well as a method for classifying the stage of the parasite based on ontology to determine the species of infection. This method is modelled after the pathologist's method of diagnosis for visual examination, hence it is anticipated that it would produce outcomes that are comparable.The Autoscope, a low-cost automated digital microscope with a set of computer vision and classification algorithms, aims to provide use cases in the developing world with accurate diagnosis of a range of infectious diseases [10]. Mohammed *et al.* [11] developed a system for processing images to recognise malaria parasites in thin blood smears and categorise them into one of the four kinds of the disease. Several methods were used during the preprocessing stage to improve the photos. The five different types of leukocytes are categorised using the genetic algorithm-based k-means clustering approach in the reduced dimensions. Microscopy is the gold standard for diagnosing malaria. When instances are found in remote rural locations, this process gets difficult because there may not be any experts available to make a diagnosis there. This issue might be resolved by automating the diagnosis procedure and using an intelligent system that can identify malaria parasites. Peñas *et al.* [12] suggested a technique that may recognise malaria parasites in thin blood smear images and detect them. Huang *et al.* [13] offered a technique for automatically identifying and detecting leukocytes. Leukocytes can generally be divided into five groups: lymphocytes, monocytes, eosinophils, basophils, and neutrophils. Different roles are played by each group in the human immune system. The primary components of a leukocyte are found in the nucleus, and this is a crucial characteristic for illness classification. Five different types of leukocytes are distinguished in this study using their nuclei. The experimental findings demonstrate that our method achieves a high and guaranteed accurate recognition rate even when only leukocyte nucleus traits are used for categorization [14]. Counting white blood cells (WBCs) gives crucial information for the diagnosis of many diseases. The use of automated counting may improve the haematological approach. The initial step of automation, segmentation, is crucial for the next stages, feature extraction and classification [15]. Despite being relatively cheap and giving the examiner the chance to count the parasites and distinguish between the various types of malaria, laboratory and clinical diagnosis have certain drawbacks.

Clinical Malaria Diagnosis

Earlier symptoms of malaria are nearly equal to symptoms of any other disease which in turn lead to overtreatment, misdiagnosis, and wrong treatment.

Laboratorial Malaria Diagnosis: Peripheral Blood Smear

Principle: By using a microscope, thick and thin blood smear images are found.

Disadvantage: Need expertise to diagnose and it requires time.

Serological Tests

Principle: These tests recognizes the antibody that works against the parasite.

Disadvantage: This method is slow and needs a lot of time to diagnose.

Quantitative Buffy Tests

Principle: Staining the blood with acridine and analyzing it under an epifluorescent microscope to find the existence of plasmodium.

Disadvantage: It can detect only *P. Falciparum*.

Rapid Diagnostic Tests

Principle: Detecting specific antigen that is produced only by malarial parasites.

Theorem: Identifying a particular antigen that is solely generated by malarial parasites.

Its inability to distinguish between various types of parasites is a drawback. Theorem: Identifying a particular antigen that is solely generated by malarial parasites.

Disadvantage: It is not able to differentiate the type of parasite Principle: Detecting specific antigen that is produced only by malarial parasites.

Disadvantage: It is cannot differentiate the type of parasite.

OBJECTIVE

The conventional malarial diagnostic tests are time-consuming and less accurate. To overcome this liability, we have designed an automatic system where in patient's blood smear image is provided and several image processing techniques are applied to the provided image. The result of this whole process is reported to the physician, which helps the physician to give the best cure to the patients based on the stage of malaria and the type of parasite obtained as the result. The other way of diagnosing is template matching, where the patient's blood smear image is compared with the affected blood smear and by comparing the blood smears, the stage of malaria and type of parasite are found.

The alternative method of diagnosis is template matching, which compares the patient's blood smear picture to the impacted blood smear in order to determine the parasite type and malarial stage.

Image Processing Techniques

Grey Scale Image Conversion

It is used for converting a 24-bit colour value to an 8-bit grey value.

Image Enhancement Techniques

These techniques are used for processing the image in such a way it is suitable for deeper analysis. Processing includes direct operation on pixels and darkening of images.

These methods are used to manipulate the picture in a manner that makes it acceptable for more thorough study.

Processing involves manipulating pixels directly and dimming pictures.

Image Filtering Techniques

These techniques are used to remove noises in the image.

Image Sharpening Techniques

These techniques are used to enhance the edges, by subtracting the un-sharp portions of the image from the original image.

Image Thresholding Techniques

It is the simplest technique for image segmentation and is used to turn a grayscale picture into a binary image.

Edge Detection: It is used for data extraction, image segmentation and for finding the boundaries of objects within images.

Region Selection

Other properties of the image are extracted using this technique. It is also used for computing the shape measurements and pixel value measurements in images.

Corner Detection

It is used for extracting the corners and for identifying the features of an image.

Gray Scale Image Conversion

It entails converting a 24-bit colour value to an 8-bit grey value in order to turn a colour picture into a grayscale image. The grey scale, which ranges from 0 to 255, reflects the pixel intensity. The process of turning the original picture into a grayscale image may be done in a variety of ways, including simple averaging, minimum averaging, maximum averaging, and weighted averaging. Selecting the right conversion technique is crucial since failure to do so would result in information loss and reduced diagnostic accuracy.

Image Enhancement Techniques

Log Transformation

This transformation is used to expand darker pixel values and simultaneously compress the higher-level values.

$$Vout = c \log (1+v) \tag{1}$$

In general, pixel's value ranges from 0 to 255, to convert it to the range 0 to 1, it is necessary to divide the pixel's value by 255.

$$Vout = \log (1 + Vin/255) \tag{2}$$

Image Negation

This technique is used for negating the pixel value. For example, if pixel value is 0, it converts it into 255. This is like complementation which makes the whiter pixels darker and the darker pixels whiter.

$$Vout = 255\text{-}VIN \tag{3}$$

Gamma Transformation

This transformation takes the pixel value and it is powered with gamma value. If gamma value is 1, the output is the same as the input. $Vout = a* (Vin) v$. Here, a is constant which is equal to 1 and Vin is the pixel value and v is the gamma value, which can be initialized. If the value of $v<1$, the image becomes whiter, if $v>1$ then the image becomes darker and if $v=1$, it remains the same.

Image Filtering Techniques

Gauss Filters

This filter uses the Gaussian function to remove the noises in the image and unwanted details and gives a blurred image.

Box Filters

Convolution filter is used, which makes use of mathematical operation to perform image filtering, wherein two arrays are multiplied to provide a third one. After filtering, every pixel would be an average of its surrounding pixel.

Min Filters

This filter replaces the centre most pixel with the darkest pixel among the neighbour pixels.

Max Filters

This filter replaces the centre most pixel with the lightest pixel among the neighbour pixels.

Image Sharpening Techniques

Image sharpening techniques are used to increase the contrast between bright and dark images to establish the features. This technique is used for better detection of edges. Initially, the difference between the pixels of the transformed image and the filtered image and the value obtained is multiplied by 2. Now the resultant value obtained after multiplying is further added to the pixels of the filtered image, which in turn will enhance the edges of the image.

Image Thresholding Techniques

This technique is used to differentiate the pixels in the image by fixing a threshold value. By using this threshold value, a malarial parasite or any other object in the binary image can be easily differentiated. Pixel values lesser than the threshold would look black and the ones whose value is greater than the threshold would look white. If we want to differentiate the object with a different colour, bitwise NOT is done with each pixel to change the colour. After applying this method, a binarized image is obtained, where each pixel value would be either 0 or 1.

Edge Detection

There are many edge detectors like the Sobel operator, the canny operator, and Robert's operator. The most commonly used edge detector is sobel operator. Here masks are used to detect the edges. Say, Hx and Hy are the masks that are used to detect if there is any image or not. Hx is an array that will move vertically to detect the presence of edges; similarly, Hy is an array that moves horizontally to detect the presence of edges. Convolution is done using a 3*3 matrix and this masked matrix, the resultant value will be placed in the centre pixel.

Region Selection

This technique identifies some of the properties of the region of images. Properties contain a list of strings separated by comma. These strings could either be 'all' or 'basic'. If it is found to be all, Region properties compute all shape measurements like Area, Bounding Box, Centroid, Convex Hull, *etc.* If it is found to be all with a grey scale image, region properties also return the pixel value measurements like Max Intensity, min Intensity, Weighted Centroid, Mean Intensity, and Pixel Values. If it is interpreted as the string 'basic', the region properties compute only the Area, Centroid and Bounding Box measurements.

Corner Detection

There are many corner detectors; the most commonly used is Harris corner detector. This is based on the local auto-correlation function of a signal which measures changes occurring locally with patches shifted by a considerable amount in different directions. Given a shift $(\Delta x, \Delta y)$ to a point (x, y), the auto-correlation function is defined as:

$$c(x, y) = \sum w[I(x_i, y_i) - I(x_i + \Delta x, y_i + \Delta y)], \qquad (4)$$

Where $I(x_i, y_i)$ represents the image function for (x_i, y_i) points in the window W centred around (x, y). $H(x, y)$ by Eigen values of $C(x, y)$. $C(x, y)$ is symmetric and positive definite that is $\alpha 1$ and $\alpha 2$ are > 0. $\alpha 1\ \alpha 2$ is equal to determinant $(C(x, y)) = AC - B2$, $\alpha 1 + \alpha 2 = \text{trace}(C(x, y)) = A + C$, Hcorner Response $= \alpha 1\ \alpha 2 - 0.04(\alpha 1 + \alpha 2)2$. It is necessary to find the corner points and estimate it as local maxima of corner response.

DESIGN OF THE SYSTEM

System Work Flow

The Malaria diagnostics system works in the following way:

1. Initially, it takes in a Giemsa-Stained image of Human Red Blood Cells as inputs.

2. Performs pre-processing to highlight the required features that are essential for the forthcoming steps.

3. Then, the system performs Image Thresholding to segregate the pixels that form the foreground (necessary) and background (unnecessary) pixels. The foreground pixels consist of all the pixels that form the RBCs in the image.

By using Watershed Algorithm, each RBC is separated and the RBCs which are infected are identified;

4. And sent to further steps to identify the severity and the stage of the parasite.

5. For each infected RBCs, edge detection and corner detection are carried out to identify the severity of the disease.

6. The system then performs template matching to identify the stage of the parasite.

7. Number of normal and abnormal RBCs are calculated and displayed.

8. Finally, testing is done for each stage to check whether the entire system works as required for all stages of the malarial parasite.

9. To enhance the experience of the users, a User Interface is designed.

The detailed working of the system is given in Section B.

Working of the System

This system makes use of OpenCV package in Python to perform operations on images. Initially, Giemsa-Stained images are read by the system by making use of OpenCV API. Operations on these images are done in such a way that it highlights the necessary features of the images. Further, the severity of the disease and the stage of the parasite is found using the techniques such as Edge Detection, Corner Detection and Template Matching. The system goes through the following stages.

Image Pre-Processing

Image pre-processing is an essential step in Image Processing as the result of the further stages highly depends on this stage. This system made use of different techniques in each step and selected the one that highly suited the given test case.

Following are the steps followed by the system during this stage as shown in Fig. (**1**).

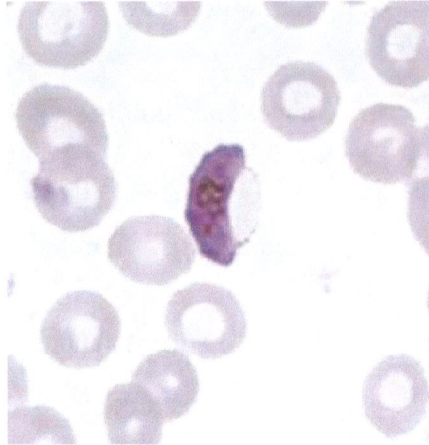

Fig. (1). Original Image.

Reading the Image

i. Initially, the input Giemsa-stained image is read by the system. The OpenCV reads the image through cv2.imread() API, where the first argument is the path of the file and the second argument is a flag denoting the way the image must be read.

ii. The format OpenCV uses to read the image is BGR (*i.e.*); each pixel in a colored image is denoted by 24 pixels, in which opencv considers the first 8 bits as Blue, next 8 as Green and so on.

Converting the Image into a Grayscale Image

i. The original image in BGR format is a 3-dimensional image, which must be converted to a Gray Scale Image (1-dimensional image), where each pixel is denoted only by its intensity value that ranges from 0 to 255. 0 denotes black, 255 denotes white and in between values denote various shades of gray.

ii. This is done because it is easier to identify the features in gray scale images rather than in colored images. Also, it reduces the complexity of the code and has better visualization as shown in Fig. (**2**).

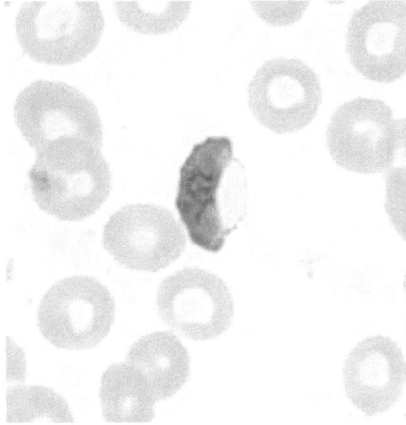

Fig. (2). Grey scale Image.

Image Transformation

Gamma Transformation: This transformation makes use of power law,

$$V\_out=a(V\_in^\gamma), \tag{5}$$

where 'a' is a constant and it is assigned the value 1. V_in, is the value of each pixel in the input Gray Scale Image that is normalized to the value between 0 to 1. γ, is the varying parameter whose value maps the intensity of each pixel of the input image to the output image. When $\gamma=1$, input image is same as that of the output, $\gamma>1$, intensity increases and in contrary $\gamma<1$, intensity decreases. In this system, γ is set to 1.5 after various trial and errors, as it produced more accurate result which is shown in Fig. (3).

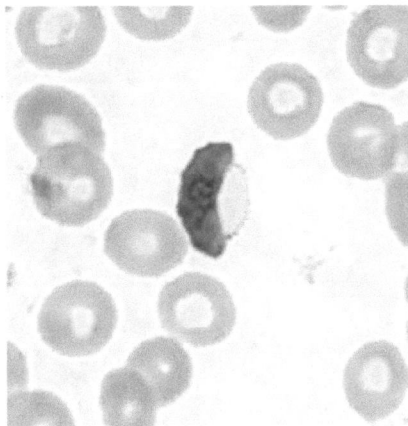

Fig. (3). Gamma Transformation.

Image Filtering

i. Gaussian Filtering: In filtering, blurring of the input image is done, to remove the noise. The input image is convolved with the Gaussian kernel matrix to produce the resultant blurred image as shown in Fig. (**4**).

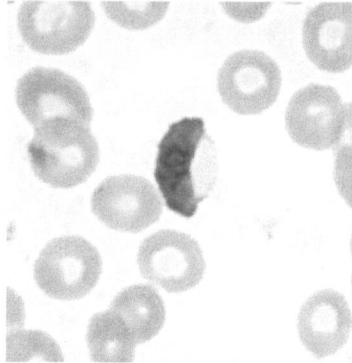

Fig. (4). Gaussian Filtering.

ii. This smoothing effect of Gaussian filter enhances edge detection and results in more accurate determination of features. The effect of this filtering in the image is evidently seen in Fig. (**4**).

Image Sharpening

i. This step is done to enhance the edges in the input image. The absolute difference of the original and the resultant image from the previous step is multiplied by a constant and added to the Gaussian filtered image as shown in Fig. (**5**).

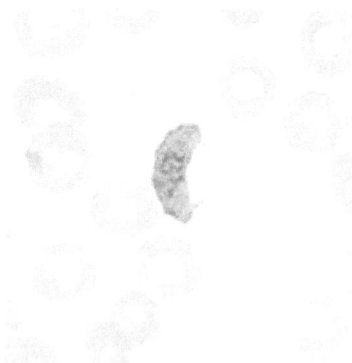

Fig. (5). Image Sharpening.

ii. The resultant image of this step is shown in Fig. (**5**).

Image Thresholding

The inference that can be obtained from the above image is that, the intensity of the parasite is lesser that RBCs which is in turn lesser than the background image. We can make use of this property to segment parasite from the above image.

RBC Segmentation

i. The resultant image from the previous stage is converted into a binary image at this stage.

ii. As each pixel value ranges from 0 to 255, the threshold value to segment RBCs is set as 225.

iii. If pixel < 225, the pixel value is set to 0 else, to 1. In this case, pixels of RBCs appear black and the background pixels appear white.

iv. The image is the complemented, so that foreground pixels appear white and background appears black.

v. The resultant binary image is passed on to the Image Segmentation class, which segments the image into individual RBCs.

vi. If there are n RBCs, each RBC is taken and again image thresholding is done to identify the parasite. Since the parasite has lesser intensity than the remaining pixels, the threshold value is chosen as 120 as shown in Fig. (**6**).

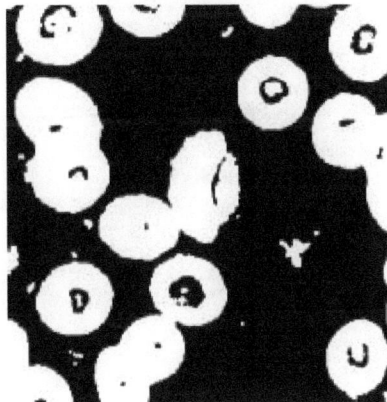

Fig. (6). Binary Image.

Steps in Segmenting Individual RBCs

In this stage, morphological operations are used to perform the segmentation of RBCs. There are four important morphological operations used in this phase.

Dilation

This process adds foreground pixels, this process works in such a way that the outer pixel is assigned the maximum value of the inner pixels.

Erosion

This process works in contrary to dilation in such a way that it removes the foreground pixels. Erosion assigns the minimum value of the inner pixel to the outer pixels.

Opening

Opening combines the fundamental operations, dilation, and erosion. The opening does the process of dilation followed by erosion.

Closing

As erosion leads to dilation, it is the process of erosion followed by dilation.

This system follows the following step to segment the image into individual RBCs is shown in Fig. (7).

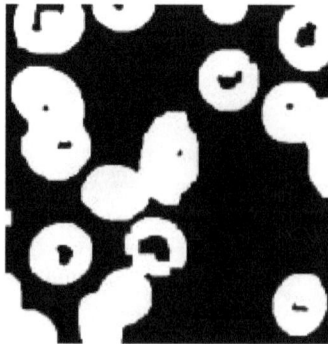

Fig. (7). Noise Removal.

i. Initially, the opening of the image is done to remove noises.

ii. In order to identify the sure foreground pixels (which are the pixels representing RBCs), the opening of the image is done for two iterations using an ellipse structuring element as RBCs are in the form of an ellipse. Pixels now in white are the sure foreground pixels as shown in Fig. (**8**).

iii. Now, in order to find the sure background pixels, closing is done. Now the pixels that are black represent the sure background as shown in Fig. (**9**).

iv. The pixels that do not fall in the category of ii and iii, are the unknown pixels as shown in Fig. (**10**).

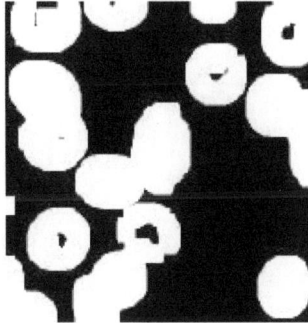

Fig. (8). Sure Foreground.

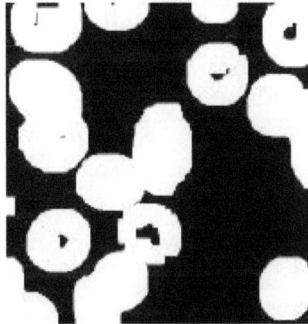

Fig. (9). Sure Background.

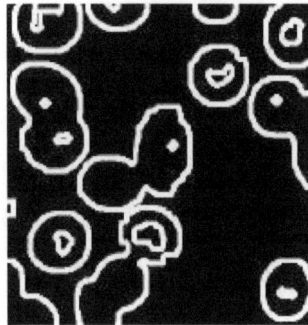

Fig. (10). Sure Background.

v. At this stage, the watershed algorithm is used on the sure foreground image, to identify the connected components. This algorithm assigns a unique numerical value to all the pixels in all directions until and unless the value differs. Consider a matrix:

$$\begin{pmatrix} 0 & 1 & 1 & 0 \\ 0 & 1 & 0 & 0 \\ 0 & 0 & 0 & 1 \end{pmatrix}$$

vi. This matrix when passed through the watershed algorithm is marked as follows:

$$\begin{pmatrix} 0 & 1 & 1 & 0 \\ 0 & 1 & 0 & 0 \\ 0 & 0 & 0 & 1 \end{pmatrix}$$

Here, as all the 0s are connected they are assigned the marker value as 0, then 1s in the first and second row are connected and hence the marker value 1 is assigned. Whereas, 1 in the third is not connected with any other one and hence is assigned a new marker value 2.

vii. In this way, each RBC in sure foreground will definitely be surrounded by background pixels and hence each RBC will be assigned different values as shown in Fig. (**11**). For each connected component, a border is applied and the impact on the original image's accuracy is as follows:

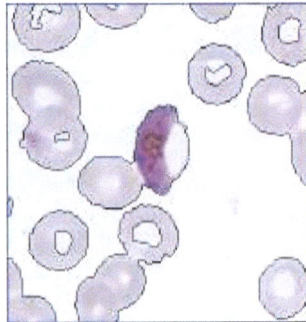

Fig. (11). Marked Image.

viii. Therefore, by using markers, we can segment individual RBCs.

ix. By setting a specific marker value to 255 and all other pixels with different marker values, we can separate each RBCs as shown in Fig. (**12**).

Fig. (12). Segmented RBCs.

x. The segmented RBCs are shown in Fig. (**12**).

Severity Calculation

At this stage, individual RBCs are interpreted to identify the parasite, and severity based on the area spread of the disease is calculated. This stage makes use of Edge detection, Corner Detection, and Area Calculation to calculate the severity. The system goes through the following steps:

i. Each segmented RBC image is passed on through the thresholding phase with a threshold value of about 120.

ii. If an RBC has pixels with intensity < 120, then it is said to be infected and the number of abnormal RBCs count is increased.

iii. Else, if there are no pixels with intensity > 120, then the RBC is normal, and the normal RBC count is increased.

For each abnormal RBC,

Edge Detection

This is carried out using Sobel Operators. This algorithm works as follows: The Sobel Operator consists of horizontal and vertical masks. Let H_x and H_y represent the horizontal and vertical masks, respectively. A horizontal mask is used to identify horizontal edges and a Vertical mask is used to identify vertical edges. Finally, these two edges are combined. The resulting edge at this stage is 2 pixels thicker and hence the outer layer of pixels must be removed to get a thinner edge. This is due to the blurring nature of the Sobel operator as shown in Fig. (**13**). The limitation mentioned above is taken care of in the Canny Edge Detection algorithm which, in the end, performs Non-maximum suppression, which is an edge-thinning technique.

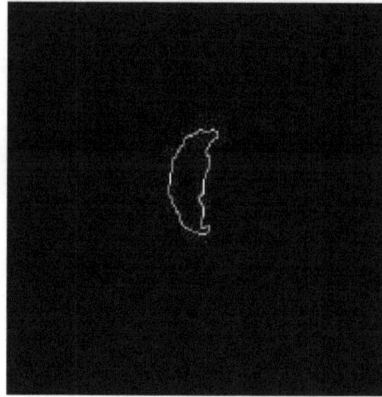

Fig. (13). Edge Detection.

Corner Detection

Corners are pixels in an image which has a large variation in intensity in all directions. Harris Corner Detection algorithm works as follows. This algorithm basically finds the difference in intensity for a displacement of (u,v) in all directions as shown in Fig. (**14**). When the difference exceeds the threshold being fixed, the pixel is marked as a corner.

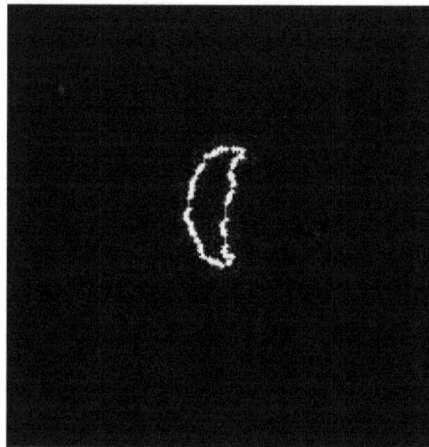

Fig. (14). Corner Detection.

iv. The number of corners found in the resultant image is stored.

Region Selection

At this stage, the area of interest (*i.e.*), the area of pixels representing the parasite, has to be calculated. This stage follows the following step:

i. Initially, assign the area as 0.

ii. The image is traversed to find the area covered by the pixels representing the parasite.

iii. The calculated value is then added to the area in step a.

iv. Likewise, all the abnormal RBCs are traversed and the aggregate of area is calculated.

Severity

Finally, severity is calculated using the formula,

$$\text{Severity \%} = 2*(\text{Corners}/\text{Area})*100 \tag{6}$$

Stage Detection

The malarial parasite goes through three major steps, Schizont, Gametocyte and Trophozite. The stage of the malarial parasite is determined by this system. In order to do so, the system makes use of the Template Matching algorithm. This algorithm follows the following step:

a. Initially, min_distance is assigned a maximum value.

b. For each template in the templates repository,

i. The template is passed on through the abnormal RBC.

ii. The distance between the template and the abnormal RBC is identified, if the value is lesser than the min_distance,

iii. Min_distance is assigned with the value in distance.

iv. Stage is set to the stage of the current template

c. Finally, the stage is displayed. The template of this stage and the parasite in the abnormal RBC have the minimum distance as shown in Fig. (**15**).

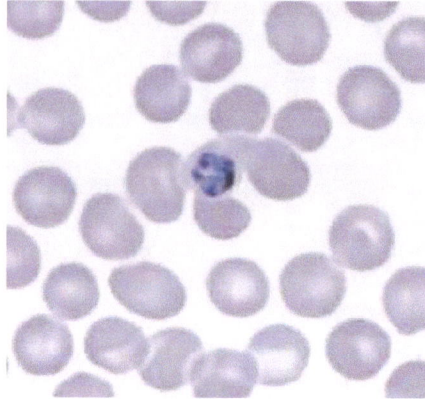

Fig. (15). Schizont stage.

CONCLUSION

Image Processing Techniques have not just made a revolution in the medical field, they have got wider applications and almost all technical fields are impacted by them. Our proposed system enables the easiest way for the physician to diagnose malaria with the highest accuracy; providing the stage of malaria, parasites affecting the blood and the severity of the disease, when the patient's blood smear image is processed. This study will be helpful to all medical organizations diagnosing malaria. Further, in this system, functionalities for finding the type of parasite that affected the blood can be identified. This as well as the stage of the parasite can be more efficiently executed using machine learning algorithms. Also, this system is not efficient if the intensity of the images has a larger difference to the training images used in this system. This is because the system uses a trial and error method to fix the threshold values and it is static. This can be rectified by finding the minimum and maximum intensity value and finding the frequency of the pixels with intensity values ranging from minimum to maximum value. And, from the gathered information, the threshold values must be calculated dynamically.

REFERENCES

[1]   A. Bashir, Z.A. Mustafa, I. Abdelhameid, and R. Ibrahem, "Detection of malaria parasites using digital image processing", *2017 International Conference on Communication, Control, Computing and Electronics Engineering (ICCCCEE),* 2017 16-18 January 2017, Khartoum, Sudan, pp. 1-5.
[http://dx.doi.org/10.1109/ICCCCEE.2017.7867644]

[2]   A. Mehrjou, T. Abbasian, and M. Izadi, "Automatic malaria diagnosis system", *2013 First RSI/ISM International Conference on Robotics and Mechatronics (ICRoM),* 2013 13-15 February 2013, Tehran, Iran, pp. 205-211.
[http://dx.doi.org/10.1109/ICRoM.2013.6510106]

[3]   W.A. Saputra, H.A. Nugroho, and A.E. Permanasari, "Toward development of automated plasmodium detection for Malaria diagnosis in thin blood smear image: An overview", *2016 International Conference on Information Technology Systems and Innovation (ICITSI),* 2016 24-27 October 2016,

Bandung, Indonesia, pp. 1-6.
[http://dx.doi.org/10.1109/ICITSI.2016.7858228]

[4]   S.K. Reni, I. Kale, and R. Morling, "Analysis of thin blood images for automated malaria diagnosis", *In 2015 E-Health and Bioengineering Conference,* 2015 19-21 November 2015, Iasi, Romania, pp. 1-4.
[http://dx.doi.org/10.1109/EHB.2015.7391597]

[5]   D. Das, M. Ghosh, C. Chakraborty, A.K. Maiti, and M. Pal, "Probabilistic prediction of malaria using morphological and textural information", *In 2011 International conference on image information processing,* 2011 03-05 November 2011, Shimla, India, pp. 1-6.
[http://dx.doi.org/10.1109/ICIIP.2011.6108879]

[6]   J.P. Vink, M. Laubscher, R. Vlutters, K. Silamut, R.J. Maude, M.U. Hasan, and G. De Haan, "An automatic vision-based malaria diagnosis system", *J. Microsc.,* vol. 250, no. 3, pp. 166-178, 2013.
[http://dx.doi.org/10.1111/jmi.12032] [PMID: 23550616]

[7]   C. Dallet, S. Kareem, and I. Kale, "Real time blood image processing application for malaria diagnosis using mobile phones", *2014 IEEE International Symposium on Circuits and Systems (ISCAS),* 2014 01-05 June 2014, Melbourne, VIC, Australia, pp. 2405-2408.
[http://dx.doi.org/10.1109/ISCAS.2014.6865657]

[8]   S.K. Reni, Automated low-cost malaria detection system in thin blood slide images using mobile phones, Doctoral dissertation, University of Westminster, 2014.

[9]   V.V. Makkapati, and R.M. Rao, "Segmentation of malaria parasites in peripheral blood smear images", *2009 IEEE International Conference on Acoustics, Speech and Signal Processing,* 2009pp. 1361-1364 19-24 April 2009, Taipei, Taiwan, pp. 1361-1346.
[http://dx.doi.org/10.1109/ICASSP.2009.4959845]

[10]   C.B. Delahunt, C. Mehanian, L. Hu, S.K. McGuire, C.R. Champlin, M.P. Horning, B.K. Wilson, and C.M. Thompon, "Automated microscopy and machine learning for expert-level malaria field diagnosis", *2015 IEEE Global Humanitarian Technology Conference (GHTC),* 2015 08-11 October 2015, Seattle, WA, USA, pp. 393-399.
[http://dx.doi.org/10.1109/GHTC.2015.7344002]

[11]   H.A. Mohammed, and I.A.M. Abdelrahman, "Detection and classification of malaria in thin blood slide images", *In 2017 international conference on communication, control, computing and electronics engineering.,* . 16-18 January 2017, Khartoum, Sudan, pp. 1-5.
[http://dx.doi.org/10.1109/ICCCCEE.2017.7866700]

[12]   K.E.D. Peñas, P.T. Rivera, and P.C. Naval, "Malaria parasite detection and species identification on thin blood smears using a convolutional neural network", *2017 IEEE/ACM International Conference on Connected Health: Applications, Systems and Engineering Technologies (CHASE),* 2017 17-19 July 2017, Philadelphia, PA, USA, pp. 1-6.

[13]   A.B. Nicholas, R.C. Nicki, R.W. Brian, and A.A.H. John, *Davidson's principles and practice of medicine.* Churchill Livingstone, 2006, p. 1381.

[14]   D.C. Huang, K.D. Hung, and Y.K. Chan, "A computer assisted method for leukocyte nucleus segmentation and recognition in blood smear images", *J. Syst. Softw.,* vol. 85, no. 9, pp. 2104-2118, 2012.
[http://dx.doi.org/10.1016/j.jss.2012.04.012]

[15]   O. Sarrafzadeh, A.M. Dehnavi, H. Rabbani, and A. Talebi, "A simple and accurate method for white blood cells segmentation using K-means algorithm", *2015 IEEE Workshop on Signal Processing Systems (SiPS),* 2015 14-16 October 2015,Hangzhou, China, pp. 1-6.
[http://dx.doi.org/10.1109/SiPS.2015.7344978]

CHAPTER 3

LMEPOP and Fuzzy Logic Based Intelligent Technique for Segmentation of Defocus Blur

R. Ravi[1,2,*], **R. Kabilan**[2], **R. Mallika Pandeeswari**[2] and **S. Shargunam**[2]

[1] *Department of CSE, Francis Xavier Engineering College, Tirunelveli, Tamil Nadu, India*

[2] *Department of Electronics and Communication Engineering, Francis Xavier Engineering College, Tirunelveli, Tamil Nadu, India*

Abstract: Defocus blur is extremely common in images captured using optical imaging systems. It may be undesirable, but may also be an intentional artistic effect, thus it can either enhance or inhibit our visual perception of the image scene. For tasks, such as image restoration and object recognition, one might want to segment a partially blurred image into blurred and non-blurred regions. In this project, we propose a sharpness metric based on the the Local maximum edge position octal pattern and a robust segmentation algorithm to separate in- and out-of-focus image regions. The proposed sharpness metric exploits the observation that most local image patches in blurry regions have significantly fewer certain local binary patterns compared with those in sharp regions. Using this metric together with image matting and multiscale fuzzy inference, this work obtained high-quality sharpness maps. Tests on hundreds of partially blurred images were used to evaluate our blur segmentation algorithm and six comparator methods. The results show that our algorithm achieves comparative segmentation results with the state of the art and has high speed advantage over others.

Keywords: Edge position, Fuzzy systems, Maximum, Octal patterns.

INTRODUCTION

Signal processing is an electrical engineering and mathematics field that focuses on the analysis and treatment of analog and digital signals, as well as signal storage, filtering, and other processes. These signals include transmit signals, audio or voice signals, image signals, and other signals. Image processing seems to be the field that deals with the forms of signals in which an image is formed

---

* **Corresponding author R. Ravi:** Department of CSE, Francis Xavier Engineering College, Tirunelveli, Tamil Nadu, India and Department of Electronics and Communication Engineering, Francis Xavier Engineering College, Tirunelveli, Tamil Nadu, India; E-mail: fxhodcse@gmail.com

S. Kannadhasan, R. Nagarajan, N. Shanmugasundaram, Jyotir Moy Chatterjee & P. Ashok (Eds.)

and the output is also an image. As the name suggests, it is concerned with image processing. Digital and analog image processing are two types of image processing.

If we can accurately recognize all of the items and their forms in a picture, it will appear crisper or more detailed. A picture of a face, for example, seems clear when we can recognize the facial components. The margins of a thing give it its form. So, when we blur the image, we just minimize the edge content and smooth out the transition from one hue to the next.

The phenomenon of defocus occurs when a picture is out of focus, reducing the sharpness as well as the contrast of the image [1 - 5]. Signal processing is a branch of electrical engineering and mathematics concerned with the examination and processing of analogue and digital signals, as well as actions such as signal storage and filtering. These signals include sound or speech communication, as well as visual messages. Image processing is the process of analysing and acting on pictures by humans.

Local sharpness assessment is the most commonly used method for defocus segmentation in the literature. In the last two decades, there have been several efforts in this area, the majority of which can be found in the field of picture quality evaluation, where images are graded by a single sharpness value that should comply with human visual perception. As a result, a blurred, high-contrast edge section could have a better sharpness rating than one that is in focus but low-contrast. These metrics work well for relative sharpness measurements, such as in focal stacking, but not for local sharpness measurements across different picture contents.

Object segmentation, dynamic compression, and object identification all benefit from the detection of visually prominent picture areas. In this research, we provide a technique for detecting salient regions that produce full-resolution representations with well-defined salient object borders. The original image's borders are retained by maintaining far more frequency of content than previous approaches [6 - 10]. This approach takes advantage of colour and brightness attributes while being easy to implement and computationally efficient. By obtaining higher precision and recall, our technique beats the five algorithms in both the ground-truth assessment and the segmentation job.

The method noise is a new metric for evaluating and comparing the overall performance of digital picture denoising algorithms. We calculate and assess the technique's noise for a broad class of denouncing methods, notably local smoothing filters, first. Second, a novel technique called the nonlocal mean (NL-means), which is based on a nonlocal average of all pixels in the picture, is used.

Finally, we compare the NL-means method with local smoothing filters in certain trials.

The local blur kernels of picture blocks are first approximated, and then the local blur levels of both the local blur kernels are measured using a blurring approach. The result of reblurring is a statistic that may be used to distinguish between blurred and nonblurred picture blocks. For the fine identification of blurry and non-blurred sections, block-based and pixel-based approaches are also used. For out now and motion-blurred photos, our method has been tested. The novel method detects and segments blurred and non-blurred areas in selective blurred images with 88 percent accuracy for normal out-of-focus blur, 86 percent accuracy for artificial out-of-focus blur, and 83 percent accuracy for artificial motion blur, outperforming state-of-the-art approaches [11 - 15].

LITERATURE SURVEY

Image Sharpness Assessment Based on Local Phase Coherence

Sharpness is a significant factor in visual picture quality evaluation. The visual processing system is capable of detecting blur and evaluating the sharpness of visual pictures with ease, but the underlying process is unknown. The majority of existing blur/sharpness evaluation techniques rely on edge energy reduction of global or local high-frequency content. Sharpness is defined as a good community of unique picture features assessed in the difficult wavelet transform domain, which is a new way of looking at the issue. In the scale space, previous LPC computations were limited to complex coefficients distributed across three successive dyadic scales. A versatile architecture is proposed that enables LPC computing in various fractional sizes. Then, without referring to the original image, make a new brightness assessment algorithm. The suggested technique was tested using four subject-rated publically available picture datasets, which showed equivalent performance when compared to state-of-the-art algorithms.

Gray-scale and Rotation Invariant Texture Classification with Local Binary Patterns

The technique is founded on the recognition that uniform local binary patterns are essential aspects of local image texture and that their occurrence histogram is a highly significant texture feature. Because the operator is stable against any monotonic change in the grey scale, the suggested technique is particularly resilient in respect of gray-scale fluctuations. Another benefit is the operator's computational simplicity. These operators define the spatial configuration for local image texture, and their performance can be enhanced by integrating them with scale-invariant variance metrics that define local image texture contrast.

These orthogonal measures' joint distributions are demonstrated to be extremely potent techniques for scale-invariant feature texture analysis.

An Image Recapture Detection Algorithm Based on Learning Dictionaries of Edge Profiles

The challenge of recovering pictures submitted with an unknown and spatially changing blur induced *via* defocus or linear motion is addressed in this study. Estimating the global picture blur is modelled as a multi-label energy reduction problem. The energy is made up of unary and binary variables that represent learning local blur estimators and blur smoothness, respectively. Ishikawa's approach, which exploits the natural balance of Gaussian blur levels for linear movements and defocus, is used to get the global minimum [8]. The picture is recovered using a robust deblurring technique based on sparse regularisation using global image statistics once the blur has been measured. We offer qualitative outcomes based on actual photos and quantitative conclusions using synthetic data. To reduce motion blur and improve the depth of field, short exposures and narrow apertures can be utilized. However, this might result in noisy photos, especially in low light [16 - 20].

PROPOSED SYSTEM

For effective implementation, the proposed work starts with somewhat blurred photos that are scaled into several orders. The LMEPOP computation is applied to the multi-scale photos. The image matting is carried out when the high contrast separation is carried out, and the sharp metric may be acquired from the LMEPOP calculation. The last phase is a fuzzy control multi-scale inference system, in which the fuzzy inferences are given three-scale pictures. The outcome of the fuzzy inference is a segmented in-focus region. The following figure depicts the proposed work's primary block diagram.

LOCAL MAXIMUM EDGE POSITION OCTAL PATTERN

Vipparthi *et al.* recently presented the notion of Local Maximum Edge Position Octal Patterns (LMEPOP), which uses the octal code generation's first three dominant edge positions. The maximum edges are detected using the Sign Maximum Edge Position Octal Pattern (SMEPOP) operator for the sign maximum edges and the Magnitude Maximum Edge Position Octal Pattern (MMEPOP) operator for the magnitude maximum edges. The LMEPOP operator manipulates three-pixel blocks in an image. Using Fig. (**1**), it first calculates the variation between greyscale values of the central pixel as well as its eight neighbors inside a 3x3 pixel block. Then the f (i) values are sorted in terms of sign and magnitude and the indexes are collected in indexs and indexm using the following equations:

$$\text{index}_s = \text{sort}\,[f(1), f(2) \dots\dots, f(8)] \tag{1}$$

$$\text{index}_m = \text{sort}\,[|f(1)|, |f(2)|, \dots, |f(8)|] \tag{2}$$

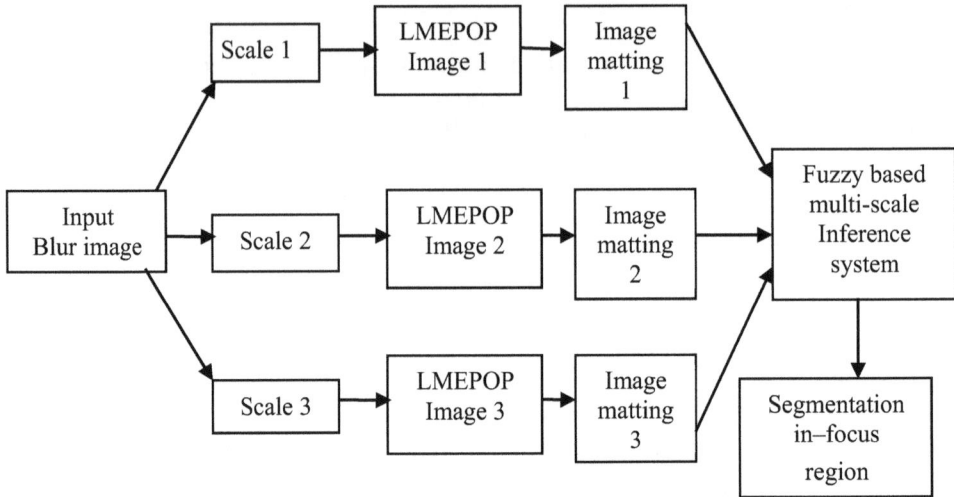

Fig. (1). Block diagram for the proposed work.

Using these indexes, the SMEPOP and MMEPOP codes are computed for the first three dominant values using the following equations:

$$\text{SMEPOP}(g_c) = 8^2 x\,\text{index}_s^1 + 8^1 x\,\text{index}_s^2 + 8^0 x\,index_s^3 \tag{3}$$

$$MMEPOP(g_c) = 8^2 x\,index_m^1 + 8^1 x\,index_m^2 + 8^0 x\,index_m^3 \tag{4}$$

Fig. (2) shows how SMEPOP and MMEPOP codes are calculated in the 35% neighborhood of a grayscale picture. The entire pixel is scanned by SMEPOP and MMEPOP histograms after each pixel's SMEPOP and MMEPOP codes have been calculated. The SMEPOP and MMEPOP histograms of a picture are 502-bin histograms because each SMEPOP and MMEPOP code has a value between 0 and 501. The image's LMEPOP descriptor is created by joining the SMEPOP and MMEPOP histograms, yielding a descriptor with 1004 bins. In addition, the research suggests constructing a local Gabor value of the maximum position octal patterns (LGMEPOP) by calculating SMEPOP and MMEPOP just on the Gabor filter response.

Grey image 3x3 Neighbourhood

65	68	63
70	67	62
66	64	69

Local Differences

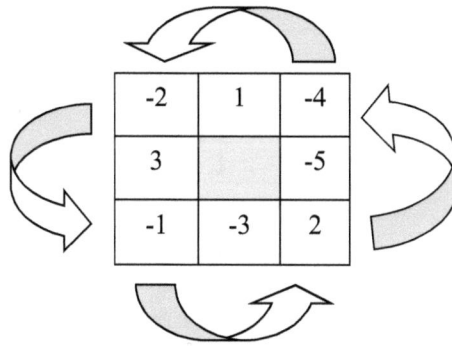

-2	1	-4
3		-5
-1	-3	2

Index and values

0	1	2	3	4	5	6	7
-5	-4	1	-2	3	-1	-3	2

SMEPOP

4	7	2	5	6	3	1	0
3	2	1	-1	-2	-3	-4	-5

$$4 \times 8^2 + 7 \times 8^1 + 2 \times 8^0 = 314$$

MMEPOP

0	1	4	3	7	6	2	5
5	4	3	3	2	2	1	1

$$0 \times 8^2 + 1 \times 8^1 + 4 \times 8^0 = 12$$

Fig. (2). Calculation of SMEPOP and MMEPOP codes.

BLUR SEGMENTATION ALGORITHM

Our algorithm for segmenting blurred or sharp patches using our LMEPOP-based sharpness metric is summarized in this section and shown in Fig. (**3**). Multi-scale sharpness map creation, alpha matting activation, alpha map calculation, and multi-scale sharpness inferences are the four primary components of the algorithm. The algorithm's output represents the inferred sharpness mapping at the largest scale. This is really a grayscale picture, and the higher the intensity, and the sharper the image.

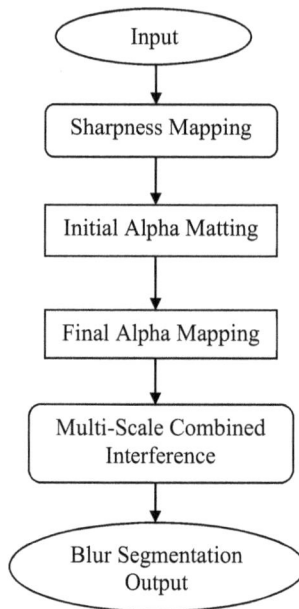

Fig. (3). Flow chart of blur segmentation.

FUZZY BASED MULTI SCALE INFERENCE SYSTEM

A fuzzy logic system (FLS) is the nonlinear mapping of such a dataset to a scalar output data set. The inference engine, rules, defuzzifier, and fuzzifier are the four basic components of an FLS. Fig. (**4**) depicts these components as well as the overall design of an FLS. Algorithm 1 outlines the situation of fuzzy logic. Using fuzzy linguistic variables, fuzzy linguistic words, and membership functions, a crisp set of input data is gathered and transformed into a fuzzy set. Fuzzification is the term for this process. Following that, an inference is constructed using a set of rules. Finally, in the defuzzification process, the fuzzy output is translated into a crisp output utilizing membership functions. In order to exemplify the usage of an FLS, consider an air conditioner system controlled by an FLS.

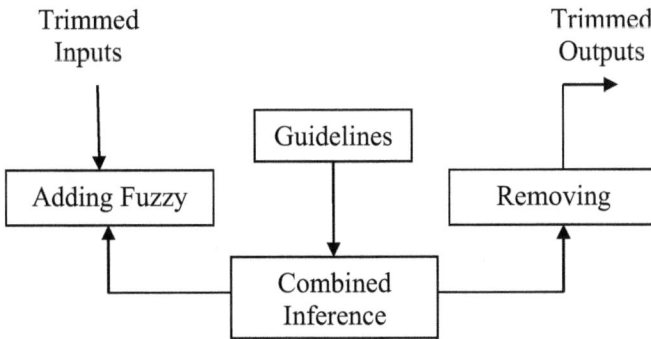

Fig. (4). Fuzzy logic controllers.

Algorithm 1 explains how it works with fuzzy logic. Utilizing fuzzified variables, fuzzy inference words, and membership functions, a clear set of input data is gathered and transformed into a fuzzy set. Fuzzification is the term for this process. Following that, an inference is constructed using a set of rules I shown in Fig. **(4)**.

Fuzzy Set Operations

Fuzzy set operations are used to evaluate the fuzzy rules as well as to combine the outcomes of the separate rules. The operations performed on fuzzy sets differ from those performed on non-fuzzy sets. The degrees of membership for fuzzy sets A and B are FA and FB, respectively. Compared to possible fuzzy procedures for OR and operators on such sets, Max and min are the most commonly used operations for OR and AND operators, respectively. For fuzzy sets, the complement (NOT) operation is employed.

Defuzzification

The classifier of such an output variable is used to defuzzify the data. Assume that at the conclusion of the inference, we obtain the result shown in Table **1**. The shaded regions in this diagram are all part of the fuzzy outcome. The goal is to extract a crisp value from this hazy outcome, which is represented by a dot in the diagram.

RESULT AND DISCUSSION

The suggested work is assessed for the proposed work using a database of 5 defocus blur photos. The example dataset photos are given in the accompanying diagram, with written commentary for the focussed and defocused regions. Sample Defocus Blur Reporter Image is taken as an input image from the database which is shown in Fig. **(5)**.

Table 1. Shows various test images.

-	Various Input Images
Teddy bear image	
Reporter image	
Goal keeper image	
Violinist image	
Old man @ park image	

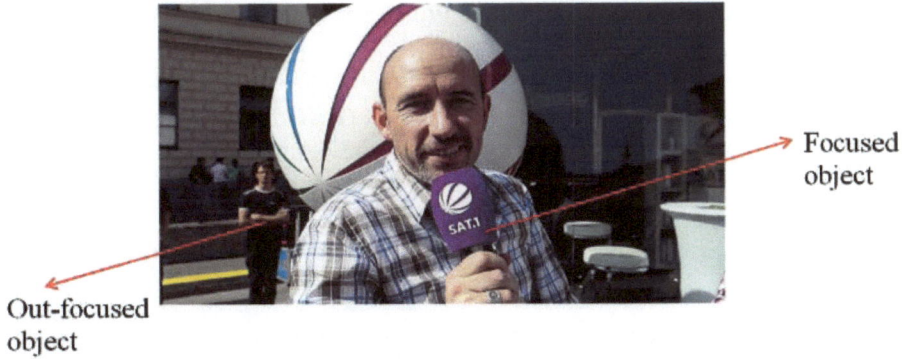

Fig. (5). Sample defocus blur reporter image.

The grey scale conversion is highly crucial for dimension reduction. Color photos are transformed into grey scale or monochrome images with the use of an inbuilt programme in MATLAB called 'rgb2gray.' The greyscale conversion of Fig. (6) is shown in the diagram below:

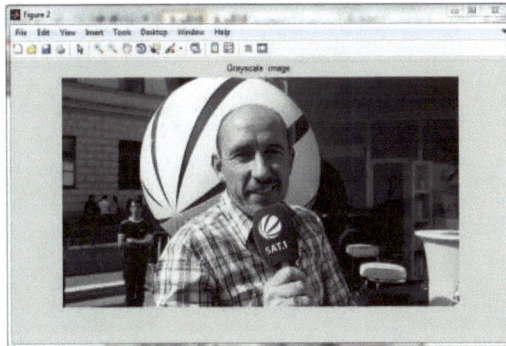

Fig. (6). Gray scale image.

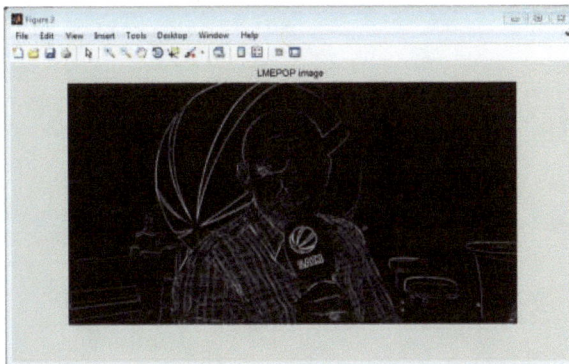

Fig. (7). LMEPOP descriptors.

The suggested local descriptors are LMEPOP shown in Fig. (**7**), which are used in this study on three distinct scales. The LMEPOP picture is shown in Fig. (**7**).

The findings for three distinct scales for the clear portrayal of focused regions are displayed in the following figures, where the sharpness matrix creates the LMEPOP pictures. The sharpness metric computed for the scales 11, 15, and 21 is shown in the following Figs. (**7**, **8** and **9**).

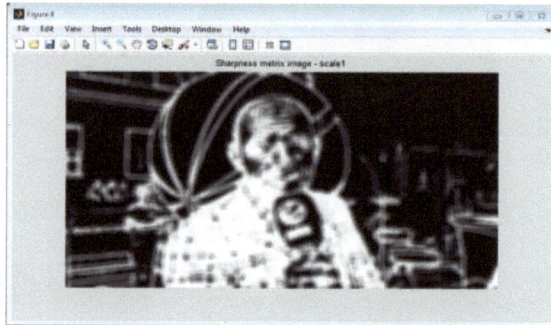

Fig. (8). Sharpness metric for scale 11.

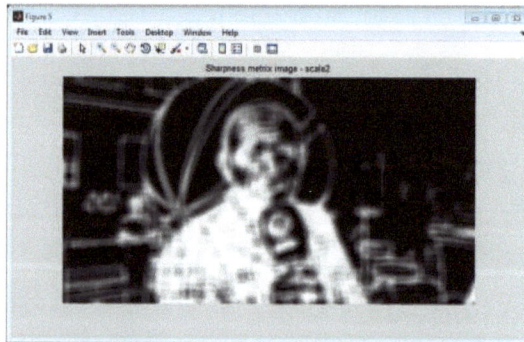

Fig. (9). Sharpness metric for scale 15.

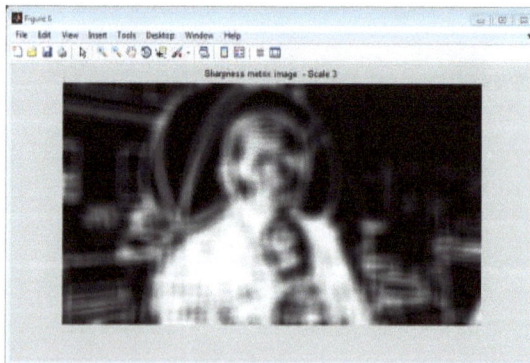

Fig. (10). Sharpness metric for scale 21.

The image matting results are retrieved from these sharpness metric pictures, and the image matting results produced in this study are shown in the Figs. (**10, 11** and **12**).

Fig. (11). Image matting results for scale 11.

Fig. (12). Image matting results for scale 15.

The picture matting results are shown for morphological procedures and then used in defocus segmentation. The result from the fuzzification system that produces the defocus blur segmentation output, in which the crispy output defines the amount of defocus, is shown in Figs. (**13** and **14**).

Fig. (13). Image matting results for scale 21.

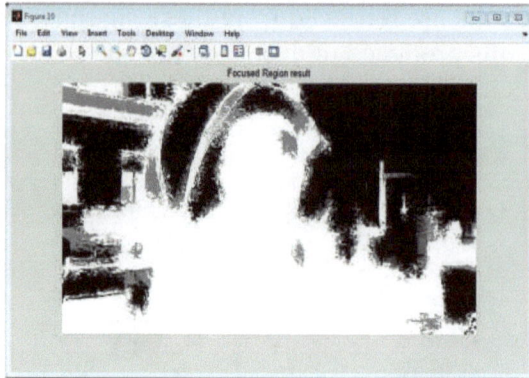

Fig. (14). Focused region segmentation output.

The defocused photographs and their related results given by the proposed method are shown by images photos in Figs. (**15** and **16**).

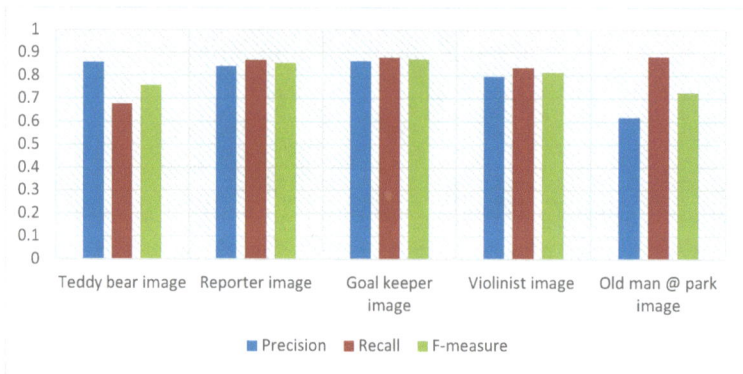

Fig. (15). Performance evaluation.

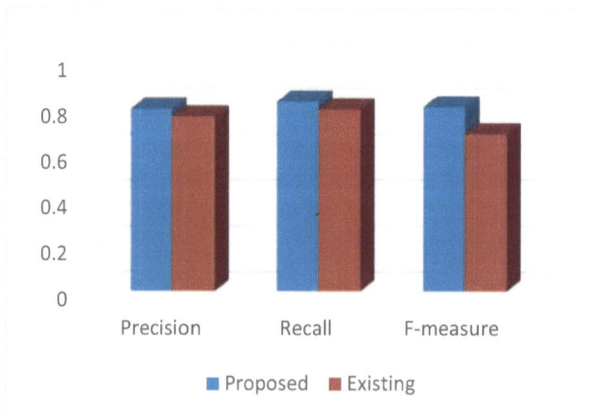

Fig. (16). Performance Comparison.

The segmentation findings were compared to the reference picture as part of the performance measurement method as shown in Figs. (**17, 18**). The validation technique yielded four values: Fake Positive (FP), Real Positive (RP), Real Negative (RN), and Fake Negative (FN) as shown in Table **2**. True positives are pixels that are correctly recognised as focused, false positives are pixels that are wrongly flagged as focused, true negatives are pixels that are correctly detected as de-focused, and Fake Negatives (FN) are pixels that are incorrectly flagged as de-focused in Table **3**.

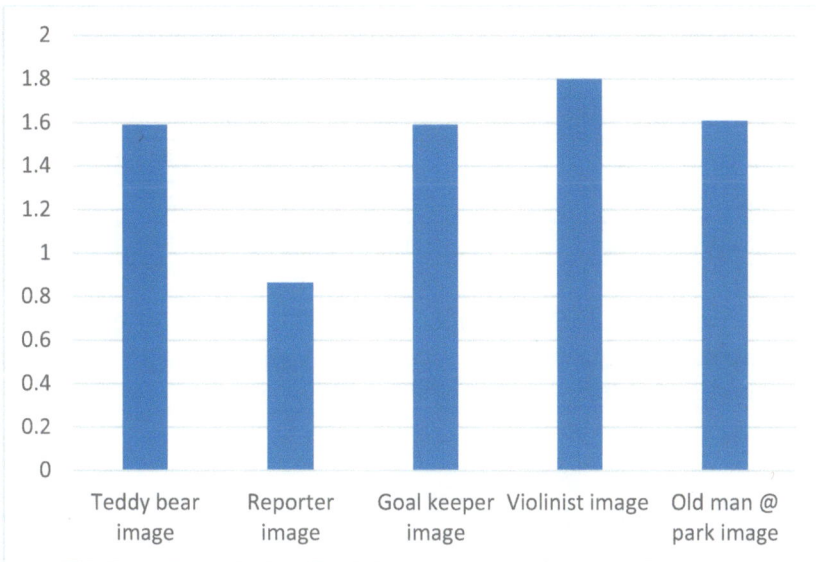

Fig. (17). LMEPOP time responses.

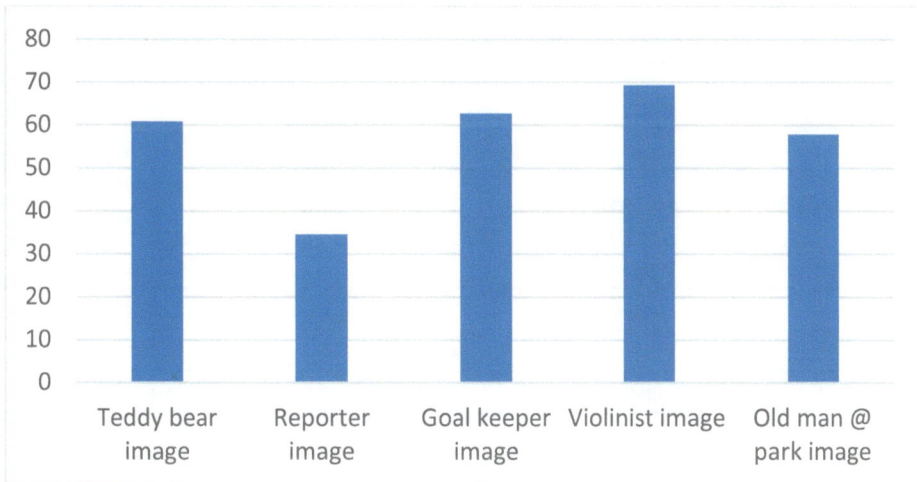

Fig. (18). Overall time taken in sec.

Table 2. Performance parameters.

-	Focused	De-Focused
Detected	Real Positive (RP)	Fake Positive (FP)
Not detected	Fake Negative (FN)	Real Negative (RN)

Table 3. Performance for parameter.

Input Images	Precision	Recall	F-measure
Teddy bear image	0.8583	0.6762	0.7564
Reporter image	0.8401	0.8654	0.8526
Goal keeper image	0.8617	0.8775	0.8695
Violinist image	0.7953	0.8322	0.8133
Old man @ park image	0.6167	0.8803	0.7253

For evaluation purpose, all the parameters are determined for each image in the dataset in Table **4**. The measurement process for various images is made using the proposed method in Table **5**.

Table 4. LMEPOP.

Image Images	LMEPOP (s)
Teddy bear image	1.59
Reporter image	0.8655
Goal keeper image	1.59
Violinist image	1.8
Old man @ park image	1.609
Average time	1.4909

Table 5. Run time for overall processing.

Image	Time Taken (s)
Teddy bear image	60.82
Reporter image	34.47
Goal keeper image	62.7
Violinist image	69.27
Old man @ park image	57.79
Average time	57.01

CONCLUSION & FUTURE ENHANCEMENT

This work proposes a defocus blur segmentation metric that is both simple and effective. The proportion of regular LMEPOP patterns in blurred and non-blurred picture areas is used to calculate this statistic. The direct usage of a local raw sharpness measurement can yield comparable results to the state-of-the-art sparse representation defocus segmentation approach, demonstrating the promise of local-based hardness measurements. It can obtain comparable outcomes to the state-of-the-art by incorporating the measure into such a multiscale data transmission framework. When utilising an autonomously and adaptively set threshold, Tseg, the algorithm's performance is maintained, according to Wit. Integral pictures can quickly implement the sharpness metric, which counts the number of specific LMEPOP patterns in a local area. The suggested approach would have a large performance advantage over previous defocus segmentation methods if paired with real-time matting techniques such as GPU versions of global matting. The multi-scale picture fusion algorithm will be implemented in the future for effective image restoration approaches.

ACKNOWLEDGEMENTS

Financial support obtained from the All India Council for Technical Education (AICTE) under Research Promotion Scheme (RPS), Sanction order no: F.No 8.9/RIFD/RPS/Policy- 1/2017-18 coordinated by Anna University Recognized Research Centre, Department of Computer Science and Engineering, Francis Xavier Engineering College, Vannarpettai, Tirunelveli 627003, Tamil Nadu, India.

REFERENCES

[1] K. Bahrami, A.C. Kot, and J. Fan, "A novel approach for partial blur detection and segmentation", *Proc. IEEE Int. Conf. Multimedia Expo (ICME)* 15-19 July 2013, San Jose, CA, USA, pp. 1-6 year 2013.
[http://dx.doi.org/10.1109/ICME.2013.6607493]

[2] T-L. Wang, K-Y. Lee, and Y-C.F. Wang, "Partial image blur detection and segmentation from a single snapshot", *2017 IEEE International Conference on Acoustics, Speech and Signal Processing (ICASSP)* 05-09 March 2017,New Orleans, LA, USA, pp.1907-1911.
[http://dx.doi.org/10.1109/ICASSP.2017.7952488]

[3] K. Purohit, A.B. Shah, and A.N. Rajagopalan, "Learning based single image blur detection and segmentation", *IEEE International Conference on Image Processing,* 2018 07-10 October 2018, Athens, Greece pp. 2202-2206.
[http://dx.doi.org/10.1109/ICIP.2018.8451765]

[4] K. He, C. Rhemann, C. Rother, X. Tang, and J. Sun, "A global sampling method for alpha matting", *in Proc. IEEE Conf. Comput. Vis. Pattern Recognit,* 2011 20-25 June 2011,Colorado Springs, CO, USA, pp. 2049–2056.
[http://dx.doi.org/10.1109/CVPR.2011.5995495]

[5] S. El-Shekheby, R.F. Abdel-Kader, and F.W. Zaki, "Restoration of spatially-varying motion-blurred images", *13th International Conference on Computer Engineering and Systems,* 2018.

[http://dx.doi.org/10.1109/ICCES.2018.8639355]

[6]   D. Krishnan, T. Tay, and R. Fergus, "Blind deconvolution using a normalized sparsity measure", *Conf. Comput. Vis. Pattern Recognit,* 2011 20-25 June 2011, Colorado Springs, CO, USA, pp. 233–240.
[http://dx.doi.org/10.1109/CVPR.2011.5995521]

[7]   R. Huang, M. Fan, Y. Xing, and Y. Zou, "Image blur classification and unintentional blur removal", *IEEE Access,* vol. 7, pp. 106327-106335, 2019.
[http://dx.doi.org/10.1109/ACCESS.2019.2932124]

[8]   F. Perazzi, P. Krahenbuhl, Y. Pritch, and A. Hornung, "Saliency filters: Contrast based filtering for salient region detection", *Conf. Comput. Vis. Pattern Recognit,* 2012 16-21 June 2012, Providence, RI, USA, pp. 733–740.
[http://dx.doi.org/10.1109/CVPR.2012.6247743]

[9]   Levin, A.; Weiss, Y.; Durand, F.; Freeman, W.T. "Efficient marginal likelihood optimization in blind deconvolution". *In Proceedings of the CVPR 2011, Colorado Springs*, CO, USA, 20–25 June 2011; pp. 2657–2664.
[http://dx.doi.org/10.1109/ACCESS.2020.2978084]

[10]   X. Tan, and B. Triggs, "Enhanced local texture feature sets for face recognition under difficult lighting conditions", *IEEE Trans. Image Process.,* vol. 19, no. 6, pp. 1635-1650, 2010.
[http://dx.doi.org/10.1109/TIP.2010.2042645] [PMID: 20172829]

[11]   T. Thongkamwitoon, H. Muammar, and P.L. Dragotti, "An image recapture detection algorithm based on learning dictionaries of edge profiles", *IEEE Trans. Inf. Forensics Security,* vol. 10, no. 5, pp. 953-968, 2015.
[http://dx.doi.org/10.1109/TIFS.2015.2392566]

[12]   F. Couzinie-Devy, J. Sun, K. Alahari, and J. Ponce, "Learning to estimate and remove non-uniform image blur", *in Proc. IEEE Conf. Comput. Vis. Pattern Recognit,* 2013 23-28 June 2013, Portland, OR, USA, pp. 1075–1082.
[http://dx.doi.org/10.1109/CVPR.2013.143]

[13]   R. Hassen, Z. Wang, and M.M.A. Salama, "Image sharpness assessment based on local phase coherence", *IEEE Trans. Image Process.,* vol. 22, no. 7, pp. 2798-2810, 2013.
[http://dx.doi.org/10.1109/TIP.2013.2251643] [PMID: 23481852]

[14]   J. Shi, L. Xu, and J. Jia, "Just noticeable defocus blur detection and estimation", *in Proc. IEEE Conf. Comput. Vis. Pattern Recognit,* 2015 07-12 June 2015, Boston, MA, USA pp. 657–665.
[http://dx.doi.org/10.1109/CVPR.2015.7298665]

[15]   N.D. Narvekar, and L.J. Karam, "A no-reference image blur metric based on the cumulative probability of blur detection (CPBD)", *IEEE Trans. Image Process.,* vol. 20, no. 9, pp. 2678-2683, 2011.
[http://dx.doi.org/10.1109/TIP.2011.2131660] [PMID: 21447451]

[16]   J. Ren, X. Jiang, and J. Yuan, "Noise-resistant local binary pattern with an embedded error-correction mechanism", *IEEE Trans. Image Process.,* vol. 22, no. 10, pp. 4049-4060, 2013.
[http://dx.doi.org/10.1109/TIP.2013.2268976] [PMID: 23797250]

[17]   H. Hu, H. Cai, Z. Ma, and W. Wang, "Semantic segmentation based on semantic edge optimization", *2021 International Conference on Electronic Information Engineering and Computer Science (EIECS)* 23-26 September 2021, Changchun, China pp.612-615.
[http://dx.doi.org/10.1109/EIECS53707.2021.9587939]

[18]   A. Shen, H. Dong, K. Wang, Y. Kong, J. Wu, and H. Shu, "Automatic extraction of blur regions on a single image based on semantic segmentation", *IEEE Access,* vol. 8, pp. 44867-44878, 2020.
[http://dx.doi.org/10.1109/ACCESS.2020.2978084]

[19]   L. Pan, Y. Dai, M. Liu, F. Porikli, and Q. Pan, "Joint stereo video deblurring, scene flow estimation and moving object segmentation", *IEEE Trans. Image Process.,* vol. 29, pp. 1748-1761, 2019.
[http://dx.doi.org/10.1109/TIP.2019.2945867] [PMID: 31613765]

[20]   T. Ojala, "Gray scale and rotation invariant texture classification with local binary patterns",
[http://dx.doi.org/10.1007/3-540-45054-8]

<h1>CHAPTER 4</h1>

Predictive Analytics - An Introduction

J. Vijayarangam[1,*], S. Kamalakannan[2] and T. Karthikeyan[3]

[1] *Department of Applied Mathematics, Sri Venkateswara College of Engineering, Sriperumbudur, Tamil Nadu, India*

[2] *Department of Information Technology, Vels Institute of Science, Technology and Advanced Studies, Chennai, India*

[3] *Department of Mathematics, Ramakrishna Mission Vivekananda College, Chennai, India*

Abstract: Analytics is one of the front runners nowadays as we have data piling in various sizes and quantities and in a dynamic fashion too. Data Analytics and in particular, predictive analytics is a hot cake in the days of social media and social networks as we grow from data banks to data rivers. This chapter is a glimpse of the basics of analytics and a few predictive analytic models currently employed in the analytical circle like Multiple regression, logistic regression and K nearest neighbor model. As we are in the era of machine learning and artificial intelligence, having a predictive analytical tool in our toolkit is all the more necessary.

Keywords: Analytics, KNN, Logistics, Regression.

INTRODUCTION

This chapter discusses predictive analytics, tools and some prevalently used models. Basically, predictive analytics forms a stage or a level in broader data analytics. Data Analytics is in general the area where we employ machine learning, statistical, mathematical models and analysis tools on data to derive valid inferences that can help us in better decision-making. It has four levels of analytics.

Descriptive Analytics

This is about what has happened in the past or about the historical data. It describes with measures like average, Dispersion, Skewness, Kurtosis or moments what exactly has happened in the system because of which we have the collected data. In other words, it describes the system through the data.

* **Corresponding author J. Vijayarangam:** Department of Applied Mathematics, Sri Venkateswara College of Engineering, Sriperumbudur, Tamil Nadu, India; E-mail: jvijayarangam75@gmail.com

S. Kannadhasan, R. Nagarajan, N. Shanmugasundaram, Jyotir Moy Chatterjee & P. Ashok (Eds.)

Diagnostic Analytics

This is the level that will tell us why, whatever we have garnered using the first level of descriptive analytics, has happened.

Predictive Analytics

This stage is used for predicting the values of one variable using one or more variables which we found out using the second level of diagnostic analytics that are related. Regression, in various forms, is the most prevalently employed predictive analytic tool as it relates quite nicely with the query we have and provides simple and clear answers.

Prescriptive Analytics

This is the most advanced kind of analytics, and it will inform us what adjustments or actions we need to make to the variables or system entities in question to get the desired results. For instance, what precise activities should a student do to acquire a 90% grade, such as reading a given text for a specified amount of hours in a specified manner [1]? It is regarded as one of the top introductions to the discipline [2 - 5]. Predictive analytics are used in [3 - 9] articles to solve a real-world issue involving consumer connections. Reference [4] is a good read on the subject of data mining, which is now a hot issue. Reference [6] will provide us with a solid foundation for comprehending knowledge domains, one of the key ideas in the use of predictive analytics. Another interesting issue, analytic language processing is well implemented in a study [7]. [8] This study is comparable to [6] but addresses a different issue: the social network, which has generated a lot of research [10]. Another research examines document clustering in the publishing domain and is an excellent study to comprehend the idea. While reference [12 - 15] are papers researching customers or customer analytics in the sector, which is a fruitful one, reference [11] is also an application paper on supply chain.

MAIN TEXT: PREDICTIVE ANALYTICS

We are going to make predictions about what could happen in the future based on what has already happened in the past—the data—as the term plainly implies. For instance, by looking at historical data from a shopping centre, we can learn more about how, what, and when customers buy. Using pattern recognition, we can then predict what, when, and how much a particular group of customers will buy. With this information, we can then make a variety of managerial decisions, such as how much to order, how much inventory we should have, how much to invest, *etc.* We use data mining, analytics, and machine learning methods for this.

Benefits and Drawbacks

Advantages

1. It is quite effective.

2. It gives us a consumer viewpoint, enabling us to serve customers more effectively.

3. We can identify dishonest consumers, brokers, or channels and stop them in order to save our resources.

4. The business's risk may be significantly decreased.

Disadvantages

1. Depends heavily on data.

2. Some techniques, such as machine learning models, may not take into consideration elements outside the data that are sometimes essential.

MODEL 1 OF PA: REGRESSION

Regression is the method used to determine the sort of relationship that exists between the variables being examined.

Depending on the situation, several definitions of "linear model" are employed. The phrase is most often used in relation to regression models, and it is commonly used interchangeably with a linear regression model. The term "linear" denotes a subclass of models that allows for a significant decrease in the complexity of the corresponding statistical theory. Let's think about the two variables X and Y. Since we are hypothetically examining their relationship, we will develop an equation that is shown in Table **1** while treating each as an independent variable.

Regression Line of X on Y [X Depending on Y]

$$X - \bar{X} = b_{xy} [Y - \bar{Y}]$$

Where,

$$\bar{X} \text{ - mean of X}$$

$$\bar{Y} \text{ - mean of Y}$$

$$b_{xy} - \text{regression coefficient of X on Y} = \frac{\sum xy}{\sum y^2}$$

$$x = X - \bar{X}$$

$$y = Y - \bar{Y}$$

Regression Line of Y on X [Y Depending on X]

$$Y - \bar{Y} = b_{xy} [X - \bar{X}]$$

Where,

$$\bar{X} - \text{mean of X}$$

$$\bar{Y} - \text{mean of Y}$$

$$b_{yx} - \text{regression coefficient of X on Y} = \frac{\sum xy}{\sum y^2}$$

$$x = X - \bar{X}$$

$$y = Y - \bar{Y}$$

Note:

1. The regression coefficients b_{xy} and b_{yx} are of the same sign.

2. The correlation coefficient and the regression coefficients are connected by:

$$r = \sqrt{[b_{xy} b_{yx}]}$$

Example 1: Calculate the regression lines for the following data.

$$X: 6\ 2\ 10\ 4\ 8$$

$$Y: 9\ 11\ 5\ 8\ 7$$

Table 1. Solution example 1.

X	Y	$x = X - \bar{X}$	$y = Y - \bar{Y}$	x^2	y^2	xy
6	9	0	1	0	1	0
2	11	-4	3	16	9	-12
10	5	4	-3	16	9	-12
4	8	-2	0	4	0	0
8	7	2	-1	4	1	-2
$\sum = 30$	$\sum = 40$	$\sum = 0$	$\sum = 0$	$\sum = 40$	$\sum = 20$	$\sum = -26$

From Table **1**, we can calculate:

$$\bar{X} = \frac{\sum X}{N} = \frac{30}{5} = 6 \; ; \bar{Y} = \frac{\sum Y}{N} = \frac{40}{5} = 8$$

Regression Coefficients:

$$b_{xy} = \frac{\sum xy}{\sum y^2} = \frac{-26}{20} = -1.3$$

$$b_{yx} = \frac{\sum xy}{\sum x^2} = \frac{-26}{40} = -0.65$$

Regression Line of X on Y [X Depending on Y]

X-6 = -1.3 [Y-8] X = -1.3Y + 1.64

Regression line of Y on X [Y depending on X]

Y-8 = -0.65 [X-6]

Y= -0.65X + 11.9

PA MODEL 2: MULTIPLE REGRESSION

Issue with regressionA multiple linear regression model is used when the research variable relies on numerous explanatory or independent factors. In two ways, this model generalizes basic linear regression. Although it does not permit arbitrary forms, it does let the mean function E(y) to rely on more than one explanatory variable and have shapes other than straight lines.

MLR -Method 1

The MLR equation can be manually calculated using the following method:

$$Y = a + b_1X_1 + b_2X_2 + ... + b_kX_k + e$$

Where,

$$b_1 = \frac{(\sum x_2^2)(\sum x_1 y) - (\sum x_1 x_2)(\sum x_2 y)}{(\sum x_1^2)(\sum x_2^2) - (\sum x_1 x_2)^2}$$

$$b_2 = \frac{(\sum x_1^2)(\sum x_2 y) - (\sum x_1 x_2)(\sum x_1 y)}{(\sum x_1^2)(\sum x_2^2) - (\sum x_1 x_2)^2}$$

$$a = \overline{Y} - b_1 \overline{X}_1 - b_2 \overline{X}_2$$

The jth regression coefficient βj is the expected change in Y per unit change in the jth explanatory variable, represented mathematically as $\beta j = dE[y]/dxj$, assuming E[error]=0.

The model is linear in parameters.

Example 2:

A person's income and education are connected. Better levels of education are often assumed to result in better income. Therefore, a simple linear regression model may be written as Income=0+1Education+.

As it is believed that even illiterate people may have some income, B1 shows the change in income relative to the change in education per unit. 0 reflects the income when education is zero.

If we add age to the list of important research variables, the regression equation will change to Income=0+1 Education+ 2 Age+.

Using a dataset and a regression equation, let's say we want to estimate the work performance of Chevy mechanics based on results from personality tests measuring conscientiousness and mechanical ability.

$$Y' = -4.10 + .09X1 + .09X2 \text{ or}$$

Job Perf = -4.10 +.09MechApt +.09 Coscientiousness.

MLR -Method 2

The multiple regression equation is:

$$\hat{y} = b_0 + b_1(x_1 - \bar{x}_1) + b_2(x_2 - \bar{x}_2) + \ldots + b_k(x_k - \bar{x}_k)$$
$$\hat{y} - b_0 = b_1(x_1 - \bar{x}_1) + b_2(x_2 - \bar{x}_2) + \cdots + b_k(x_k - \bar{x}_k)$$
$$\hat{y} - b_0 = \sum_{j=1}^{k} b_j(x_j - \bar{x}_j)$$

Given a set of n points $(x_{11}, \ldots, x_1k, y_1), \ldots, (x_{n1}, \ldots, x_{nk}, y_n)$, our objective is to find a line of the above form which best fits the points. As in the simple regression case, this means finding the values of b_j coefficients for which the sum of the squares, expressed as follows, is minimum:

$$\sum_{i=1}^{n} (\hat{y}_i - y_i)^2$$

The best line of fit is:

$$\hat{y} - \bar{y} = \sum_{j=1}^{k} b_j(x_j - \bar{x}_j)$$

Where the coefficients are given by the solutions of:

$$cov(y, x_j) = \sum_{m=1}^{k} b_m \cdot cov(x_m, x_j)$$

For two indpt variables, we have:

$$cov(y, x_1) = b_1 cov(x_1, x_1) + b_2 cov(x_2, x_1)$$
$$cov(y, x_2) = b_1 cov(x_1, x_2) + b_2 cov(x_2, x_2)$$

For example, if we have "Price" of a product, predicted using two explanatory variables- "Color" and "Quality", we might use:

Using the covariance equations:

$$cov(y, x_1) = b_1 cov(x_1, x_1) + b_2 cov(x_2, x_1)$$
$$cov(y, x_2) = b_1 cov(x_1, x_2) + b_2 cov(x_2, x_2)$$

on some data to yield equations like,

$$20.5 = 5.80b_1 - 2.10b_2 \quad 15.35 = -2.10b_1 + 6.82b_2$$

Solving, we get:

$$b_1 = 4.90 \text{ and } b_2 = 3.76.$$

Thus, the regression line takes the form:

$$(y - \bar{y}) = b_1(x_1 - \bar{x}_1) + b_2(x_2 - \bar{x}_2)$$

Using the means found in Fig. (**1**), the regression line for Example 1 is:

$$(\text{Price} - 47.18) = 4.90 \,(\text{Color} - 6.00) + 3.76 \,(\text{Quality} - 4.27)$$

or equivalently,

$$\text{Price} = 4.90 \cdot \text{Color} + 3.76 \cdot \text{Quality} + 1.75$$

Thus, the coefficients are:

$$b_0 = 1.75, \, b_1 = 4.90 \text{ and } b_2 = 3.76.$$

PA MODEL 3: LOGISTIC REGRESSION

Finding the kind of connection between a collection of independent (explanatory) factors and a categorical dependent variable (men and women) is made possible by the use of logistic regression. When the dependent variable has just two values, such as 1 and 0, we often utilise logistic regression. In contrast to multiple regression, where we forecast the value of a continuous variable, in logistic regression, we predict a logit transformation of the dependent categorical variable using a collection of explanatory factors and a mathematically sound model. Typically, we use 0 to represent the negative side and 1 to represent the good aspect. 1-p is the probability of an outcome of 0 if p is the fraction of observations having a result of 1. The odds are denoted by the ratio p/(1-p), and the logitis the odds' logarithm, or simply log odds. The logit transformation is expressed mathematically as $l=logit(p)=ln(p/(1-p))$. Using the following Fig. **(1)**, we can compare with linear regression.

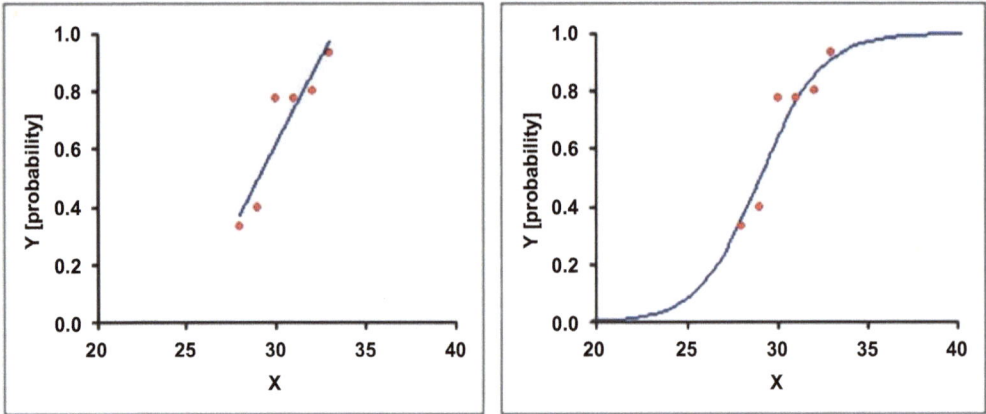

Fig. (1). Logistic curve and linear curve comparison.

PA MODEL 4: KNN

A classification model based on a distance metric is known as the KNN or K-nearest neighbor model. This categorization model uses supervised machine learning. Many refer to it as a "Lazylearning" algorithm because of the way it operates. For a given k number, it essentially assigns a data point to a class based on the classes of its nearby items in the dataset. It goes without saying that having distinct classes for various k values is a significant drawback. On the plus side, it offers a number of features, including a straightforward and very accurate method and the fact that it makes no assumptions about the data's distribution.

Algorithm

First, choose a number k for each new instance.

Step 2: To determine the k closest ones, measure the distance between the new instance and the data that has been saved.

Step 3: Choose the most popular class from the k closest ones you've chosen, then assign it to the new instance.

The algorithm's hard element is choosing k, thus we use alternatives like k=sqrt(n) or k=sqrt(n)/2, where n is the amount of the data.

Example 3: Suppose we wanted to determine a person's t-shirt size based on his height and weight, using the following information from Table **2**.

Table 2. Data -example 3.

Height	Weight	T-Shirt Size	Distance
158	58	M	4.242641
158	59	M	3.605551
158	63	M	3.605551
160	59	M	2.236068
160	60	M	1.414214
163	60	M	2.236068
163	61	M	2
160	64	L	3.162278
163	64	L	3.605551
165	61	L	4
165	65	L	5.656854
168	65	L	8.062258
168	62	L	7.071068
168	66	L	8.602325
170	63	L	9.219544
170	64	L	9.486833
170	68	L	11.40175
161	61	?	-

From Fig. (**2**), we can deduce that the triangle is the element to classify and the blue (Medium) and red (Large) ones are the data set elements from two classes. As there are more blues around the triangle, we'll assign the new element to the blue class or the medium ones.

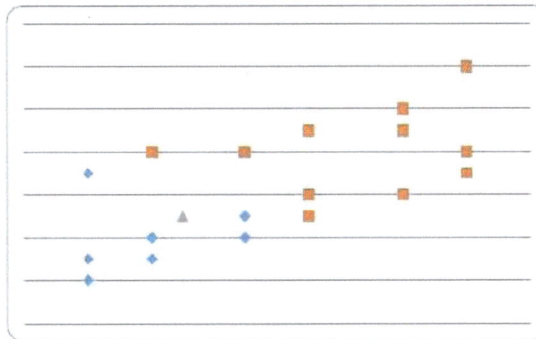

Fig. (2). Scatter plot- example 3 dataset.

CONCLUSION

We can clearly infer from the discussion above that predictive analytics is one of the most fascinating and in-demand areas of analytics, and that we need these models to assist us in making some key management choices for our company. The key benefit is the availability of a large number of models, with regression itself providing a wide range of options. Additionally, the implementation of these models on software programmes like R, Python, and Sas makes it much simpler, more current, and a scalable alternative since we don't have to worry about things like data amount, data type, or data pretreatment. Only a good and precise issue description and a reliable and trustworthy data collection method are necessary to launch the analytics. The platforms and models will take care of the rest.

REFERENCES

[1]   M. Tom, *Machine learning.* McGraw, 1997, pp. 0-7.

[2]   C. Bishop, *Pattern Recognition and Machine Learning* Springer-Verlag: Berlin, 2006.

[3]   T Mirzaei, and L Iyer, "Application of predictive analytics in customer relationship management: A literature review and classification", *A literature review and classification. InProceedings of the Southern Association for Information Systems Conference,* 2014, pp. 1-7.

[4]   D. Mimno, "Computational historiography: Data mining in a century of classics journals", *ACM J. Comput. Cult. Herit.,* vol. 5, no. 1, pp. 1-19, 2012.
[http://dx.doi.org/10.1145/2160165.2160168]

[5]   J. Banks, J.S. Carson, and B.L. Nelson, *Discrete event system simulation.* 5th. Pearson, 2010.

[6]   K. Borner, C. Chen, and K.W. Boyack, "Visualizing knowledge domains", *Ann. Rev. Inf. Sci. Technol.,* vol. 37, no. 1, pp. 179-255, 2003.

[7]   D. Hall, D. Jurafsky, and C.D. Manning, "Studying the history of ideas using topic models", *Proceedings of the Conference on Empirical Methods in Natural Language Processing (EMNLP'08),* Honolulu, Hawaii, Association for Computational Linguistics, pp. 363–371, 2008.

[8]   L. Freeman, "Visualizing social networks", *J. Soc. Struct.,* vol. 1, no. 1, 2000.

[9]   E.W.T. Ngai, L. Xiu, and D.C.K. Chau, "Application of data mining techniques in customer relationship management: A literature review and classification", *Expert Syst. Appl.,* vol. 36, no. 2, pp. 2592-2602, 2009.
[http://dx.doi.org/10.1016/j.eswa.2008.02.021]

[10]   Y. Peng, G. Kou, Z. Chen, and Y. Shi, "Recent trends in data mining (dm): Document clustering of dm publications", *Conference on Service Systems and Service Management,* 25-27 October 2006, Troyes, France, Vol. 2, pp. 1653-1659.

[11]   T. Schoenherr, and C. Speier-Pero, "Data science, predictiveanalytics, and big data in supply chain management: Current state and future potential", *J. Bus. Logist.,* vol. 36, no. 1, pp. 120-132, 2003.

[12]   E. Barkin, "CRM + Predictive Analytics", *Why It All Adds Up. CRM Magazine,* vol. 15, p. 5, 2011.

[13]   W. Buckinx, and D. Van den Poel, "Customer base analysis: Partial defection of behaviourally loyal clients in a non-contractual FMCG retail setting", *Eur. J. Oper. Res.,* vol. 164, no. 1, pp. 252-268, 2005.
[http://dx.doi.org/10.1016/j.ejor.2003.12.010]

[14]   W. Buckinx, G. Verstraeten, and D. Van den Poel, "Predicting customer loyalty using the internal

transactional database", *Expert Syst. Appl.,* vol. 32, no. 1, pp. 125-134, 2007. [http://dx.doi.org/10.1016/j.eswa.2005.11.004]

[15]   J. Burez, and D. Van den Poel, "CRM at a pay-TV company: Using analytical models to reduce customer attrition by targeted marketing for subscription services", *Expert Systems with Applications,* vol. 32, no. 2, pp. 277-288, 2007.

CHAPTER 5

Discrete Event System Simulation

J. Vijayarangam[1,*], **S. Kamalakannan**[2] and **R. Sebasthi Priya**[3]

[1] *Department of Applied Mathematics, Sri Venkateswara College of Engineering, Sriperumbudur, Tamil Nadu, India*

[2] *Department of Information Technology, Vels Institute of Science, Technology and Advanced Studies, Chennai, India*

[3] *Department of Mathematics, University College of Engineering, Tiruchirappalli, India*

Abstract: Simulated systems are used for modelling and analysis of systems for which an analytical solution is either not accessible or difficult to achieve. Simulation is a highly flexible and adaptable discipline within computer science. Because it is simpler than conventional approaches, which are often challenging, simulation is also chosen as a method of system analysis. Because of this, simulation is an area with extensive application and demand, making it interesting and beneficial to have a chapter dedicated to researching simulation with a case study of modelling a Queuing system.

Keywords: Multi-server, Queuing Model, Simulation.

INTRODUCTION

This chapter provides a basic discussion on simulation, the areas of application and one case study. Simulation is a step that a modeler or a researcher turns to when he cannot find any solution for a problem analytically. This is considered a multi-disciplinary heuristic line as many researchers from various fields have contributed to the models and tools in simulation. This is applied to a wide area of fields like production, scheduling, logistics, *etc*.

Simulation is applicable in almost all imaginable fields, of course with the restriction for a proper requirement and a decent approach in applying it. Throughout history, it is applied in manufacturing systems, public systems of many disciplines, transportation systems, construction systems, *etc*.

In manufacturing systems, it is applied in material handling areas, inventory areas, assembly sections, scheduling areas, product-mix decisions, *etc*.

* **Corresponding author J. Vijayarangam:** Department of Applied Mathematics, Sri Venkateswara College of Engineering, Sriperumbudur, Tamil Nadu, India; E-mail: jvijayarangam75@gmail.com

In the public systems, we can use it in health care, military, waste management, power plant areas, and oilspilling modeling.

In the construction area, we apply it for applications in earthmoving, strip-mining, cable-stayed bridges, strengthening the design, advanced project planning paradigm, *etc.*

In transportation areas, we apply it for cargo transfer, personnel launch systems, container operations, tollplaza operations, *etc.*

We can also use it in restaurants, entertainment, food processing, computer system performances. References [1 - 3] are excellent resources for learning about discrete event simulation or simulation in general. References [4] and [5, 6] are books of a similar kind that may be used to study queuing theory. It is a discrete event simulation application paper for queueing systems. References [7 - 10] are comparable articles that may provide us with a thorough understanding of how the toic is used in practise. A great work on the use of simulation in the economic realm can be found in [11], while another article on the use of simulation in healthcare can be found in [12].

MAIN TEXT: SIMULATION

Simulating a system's activities allows us to better understand its behaviour and implement the findings in the system. In order to make conclusions about the system, we are thereby creating a fake history of the system. The model is the imitation. Simulation may be used to investigate systems that are still in the design phase and are not yet operational. Therefore, either we are modelling a hypothetical system to determine how to implement it or we are examining an actual system to see how modifications to it would affect it.

A complicated system's internal interactions may be studied *via* simulation. Separate studies may be done to examine how changes affect different subsystems. We can learn more about crucial system parts by undertaking sensitivity analysis. Simulated data may be used to validate solutions as well.

Advantages and Disadvantages

The benefits of simulation are many. Since it replicates the system, it mostly appeals to clients intuitively. It is possible to investigate new guidelines. Both organisational and informational fluxes may be recognised. You may build new layouts and designs. It is possible to identify theories about certain phenomena. It is possible to learn how a system operates exactly.

Simulating situations have several drawbacks. Building models is a difficult procedure that requires professionals. As most of the outputs are random variables, their interpretation may be challenging. The whole procedure may be costly and time-consuming. We often employ it even though there are analytical answers because of its popularity.

System in a Simulation

A system is a collection of connected items that cooperate to achieve a single objective. In order to simulate anything, we must first identify the system. Its environment refers to the things outside the system that could have an impact on operations within the system. So, a system's border is the line separating it from its surroundings.

Any item of interest in a system is considered an entity. The traits of the entities are referred to as their personalities. A time period of a certain duration is referred to as an activity. A system state is a group of things that at any one moment characterise the system. An activity is any instantaneous event that modifies the state of the system. Exogenous events and activities are those that take place outside the system. Endogenous events and activities are those that take place within the system.

These system types may be generally categorised:

• Systems classified as discrete only experience discrete state changes.

• Continuous systems are those whose state is always changing.

Model of a System

A model is a representation of a system used to learn more about it. We include the characteristics of a system that interests us in a model. We model a system in part because it is difficult to examine the system directly, we are unable to make modifications to an active system, or the system doesn't exist in real life. We extensively examine the models to learn about the system and then apply those facts to the system to increase its effectiveness.

Models may roughly be divided into mathematical and physical categories. Symbolic notations and mathematical equations are used in mathematical models to depict systems. Physical models attempt to capture a system's underlying physics.

(i) Static: Models which represent systems only in discrete points of time.

(ii) Dynamic: Models which represent systems continuously.

(iii) Deterministic: models, which have all the input variables as deterministic.

(iv) Stochastic: Models, which have at least one input variable as random.

(v) Discrete: Models, which represent discrete systems.

(vi) Continuous: Models, which represent continuous systems.

Steps in a Simulation Study

1. Problem definition

2. Setting of objectives and project plan

3. Model building

4. Data collection

5. Model translation

6. Verification

7. Validation

8. Experimental design

9. Production runs & analysis

10. Decision of no. of runs

11. Documentation & reporting

12. Implementation (here the analyzed results of the model are implemented in the real system. Once again the analyst has the responsibility of translating the results properly to the user.

CASE STUDY: A TWO SERVER QUEUING SYSTEM

Problem Description

Simulate a Two server Queuing system for a fixed number of minutes using the R platform and derive basic measures to understand the working nature of the system.

A Queuing system, Fig. (**1**), is made up of the following aspects:

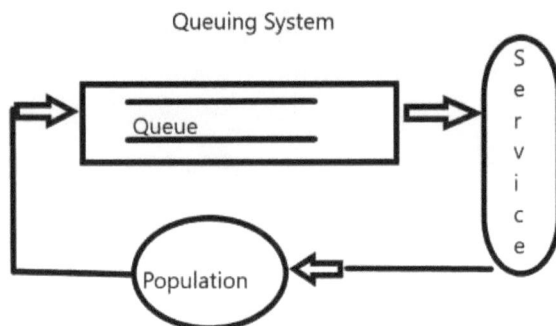

Fig. (1). Queuing System.

Queue: Any waiting line.

Queuing Models: Depicts theoretically the activities of a queuing system.

Server: Anything which provides service.

Customers: Anything which receives service.

Purpose of Queuing Models: Obtain Two Aspects of a Queueing System

1. The average waiting time of a customer.

2. The average idle time of a server with other relative measures.

Characteristics of a Queuing System

Customers arrive from the calling population to the system, wait if necessary in the queue, get service from the server, leave the system, and join the calling population to once again become potential customers. It normally possesses the following attributes as shown in Fig. (**1**).

Arrival

Customers arrive at the system, either deterministically or at random. If it is random, we go for exponential distribution to represent the arrival pattern of customers.

Service

The time taken to serve customers can be deterministic or random. Random service times usually are represented by negative exponential distributions. The number of servers can be one or many, series or parallel or mixed. Parallel servers provide service simultaneously. Series service is provided to a customer by many servers together to service him. We also can have a combination of both.

System Capacity

This is the number of customers our system can hold at any time. This can be finite or infinite.

Size of Calling Population

This is the size of the population from which the customers come. This can be finite or infinite.

Queue Discipline

This refers to the mechanism employed to select customers from the queue to provide service. This can be FIFO, LIFO, SIRO, *etc.*

Human Behavior

This refers to the behavior of a customer, if he is human. It can fall into any one of the following three types.

i. BALKING: Leaving the system without entering it.

ii. RENEGING: Leave the queue after waiting some time without getting service.

iii. JOCKEYING: To move from one queue to another.

Based on the above disciplines, we can represent a parallel queueing system using KENOALL'S Notation as a|b|cd|e|f.

a -> Inter-arrival time distribution.

b -> Service time distribution.

c -> Number of parallel servers.

d -> System capacity.

e -> Size of calling population.

f -> Queue discipline.

Measures of Queues

To understand the characters of queues, we need measures. We broadly classify them into two types:

1. TRANSIENT: Time-dependent measures.

2. STEADY-STATE: Time-independent (or) stabilized measures.

Generally, we have 5 measures which help us solve many things in a system.

P -> The server utilization.

L -> The number of customers in the system.

L_Q -> The number of customers in the queue.

W -> The average time spent by a customer in the system.

W_Q -> The average time spent by customers in queue.

These have a parallel set of measures in the transient group, based on the simulated time. When we perform a simulation, get a measure based on that run, it is transient. When we allow it to stabilize and get a measure, it is steady state. So they are also called long-run measures of the system.

An important result in these applications is the conservation equation (or) Little'sequation.

$$L = \lambda W$$

which connects three measures. The main point is $W_Q = 1/\mu$ and $L_Q = \lambda W_Q$, making it a single equation connecting 5 measures. So it is an important tool in the measuring aspect of queueing systems.

Theoretical Two server Queues: M|M|c

This model has,

Markovian arrival,

Markovian service,

Multiple server (c),

Default (System capacity

Calling population

Queue discipline)

The arrival rate is λ; the Service rate is μ. Server Utilization or System Load is:

$$\rho = \lambda / C\mu$$

Primary long-run measures of performance are:

• L long-run time-average number of customers in the system.

• L Q long-run time-average number of customers in queue.

• W long-run average time spent in the system per customer.

• w Q long-run average time spent in queue per customer.

• ρ server utilization.

[1] gives a nice and compact representation of the mathematical formulas.

We will simulate a two-server Queuing system with some basic assumptions regarding the parameters, the arrival rate, and the service rate. Derive the measures. Also, calculate the measures using the formulas provided by an analytical way, compare and conclude about simulation, (Table **1**).

Table 1. Comparison of queuing measures using theoretical model and simulation.

-	lambda	mu	c	R0	P0	Lq	Wq	L	W
Theoretical	2	3	2	0.3333	0.5	0.1	0.05	0.753	0.376
Simulation	2	3	2	0.33	0.5	0.074	0.037	0.74	0.37

CONCLUSION

We can easily see from the description above that simulation is one of the great ways to examine a system by modelling it in a straightforward, heuristic fashion. The main problem is likely resilience as the system becomes more complicated. The modeler's ability to manage such a system will be crucial in this situation since the simulation is similar to a computer in that it operates on the principle of "garbage in, garbage out." Therefore, we must exercise caution when simulating systems for which there is no previous theoretical research, at least in basic models, since we lack a suitable reference point in such situations. In addition to this, simulation is a lovely method of system analysis. Because we often start a research study in a state of chaos and simulation may provide us with the appropriate confidence-boosting start for the research study, we can utilise simulation as a suitable beginning point of the analysis if we are concerned about this constraint.

REFERENCES

[1]   J. Banks, J.S. Carson, and B.L. Nelson, *Discrete event system simulation.* 2nd. Prentice-Hall: Upper Saddle River., 2010.

[2]   R.B. Cooper, *Introduction to queueing theory.* North Holland, 1981, pp. 119-122.

[3]   A.M. Law, and W.D. Kelton, *Simulation study and analysis.* McGraw Hill Book Co: NY, U.S.A., 1991.

[4]   Z.L. Joel, "Discrete–event simulation of queuing systems", *Proceedings of the Sixth Youth Science Conference,* 2000.

[5]   M. Ehsanifar, N. Hamta, and M. Hemes, "A simulation approach to evaluate performance indices of fuzzy exponential queuing system (An M/M/C Model in a banking case study", *J. Ind. Eng. Manag.,* vol. 4, no. 2, pp. 35-51, 2017.

[6]   S. Kariuki Mwangi, "An empirical analysis of queuing model and queuing behaviour in relation to customer satisfaction at jkuat students finance office", *Am. J. Theor. Appl. Stat.,* vol. 4, no. 4, pp. 233-246, 2015.
        [http://dx.doi.org/10.11648/j.ajtas.20150404.12]

[7]   M. M Kembe, E. S Onah, and S Iorkegh, "A study of waiting and service costs of a multiserver queuing model in a specialist hospital", *Int. J. Sci. Technol. Res.,* vol. 1, pp. 19-23, 2012.

[8]   Z. Jian, "Carried out a simulation for a hospital clinic.queue system", *Systems Engineering-Theory& Practice,* vol. 3, pp. 140-144, 1998.

[9]   Wang Jun, and Qiaoying Yu, *J. Syst. Simul.,* vol. 2, pp. 574-580, 2006.

[10]   Q. Wu, *Application Research of Computers,* vol. 4, pp. 43-45, 1995.

[11]   N. Azmat, *Department of Economics and Society, M. Sc.* University of Dalarna: Hogloskan Dalarna, Sweden, 2007.

[12]   A. Saxen, "Accident and Emergency Section Simulation.in.Hospital", Available from: www.wseas.us/e-library/conferences/digest2003/papers/466 124.pdf

CHAPTER 6

Performance Analysis of Different Hypervisors Using Memory and Workloads in OS Virtualization

J. Mary Ramya Poovizhi[1,*] and **R. Devi**[1]

[1] *Department of Computer Science, Vels Institute of Science, Technology and Advanced Studies (VISTAS), (Deemed to be University), Chennai, India*

Abstract: Virtualization is a cloud-computing technology that only needs one CPU to work. Virtualization makes it look like many machines are working together. Virtualization focuses mainly on efficiency and performance-related tasks because it saves time. This paper primarily focuses on operating system virtualization. It is the modified form of a standard operating system that allows users to operate different applications that produce a virtual environment to perform various tasks on the same machine by running other platforms. This virtual machine helps compare the performance of Type1 and Type2 hypervisors based on how much work they do and how much memory they use.

Keywords: Bare metal hypervisor, Cloud computing, Deadline, Deployment model, Energy consumption, Hosted hypervisor, Hyper-V, KVM, Optimization, OS virtualization, Service model, Task scheduling, Virtualized cloud, Vm placement, VMware, Xen.

INTRODUCTION

Cloud computing is a way to run applications and let people pay for their use. It allows people to use server networks and pays for what they use. Cloud computing has a lot of significant advantages, like flexible and scalable infrastructures, lower operational and maintenance costs, and more high-performance applications available [1]. Virtualization means that it is possible to run one or more virtual machines on the same computer simultaneously. Each virtual machine has its own (guest) operating system (OS) [2]. In Virtualization, there are many different ways to make things work. Everything you need to run your computer and get your data is here.

---

[*] **Corresponding author J. Mary Ramya Poovizhi:** Department of Computer Science, Vels Institute of Science, Technology and Advanced Studies (VISTAS), (Deemed to be University), Chennai, India;
E-mail: devi.scs@velsuniv.ac.in

S. Kannadhasan, R. Nagarajan, N. Shanmugasundaram, Jyotir Moy Chatterjee & P. Ashok (Eds.)

A hypervisor is a process or a thing that does something. Administrators can keep operating systems and applications separate from the hardware that makes them work when administrators can do this. People who use cloud computing the most use virtual machines, or VMs, because they let multiple guest operating systems (also called virtual machines or VMs) run simultaneously on the same host computer, which is called a host computer. People who run computers can use many resources more efficiently if they split up computing resources (RAM, CPU, and so on) between multiple virtual machines (VM). (Fig. **1**).

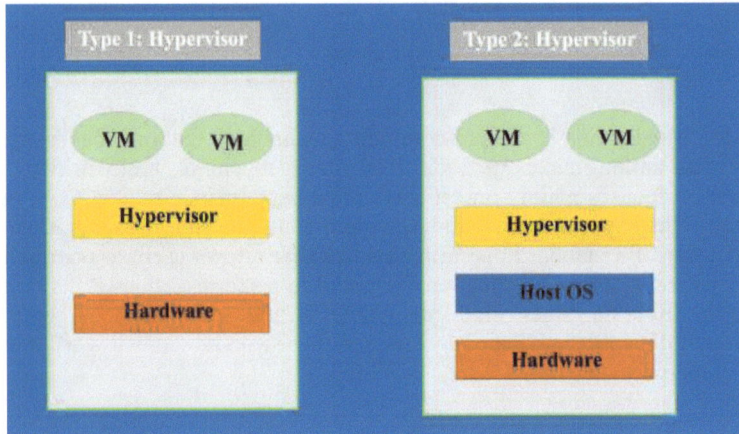

Fig. (1). Types of Hypervisors.

In virtualization, a computer system's virtual instance is created in an isolated layer from its authentic hardware. Virtualization enables a computer system to run multiple operating systems simultaneously. It is also possible for applications running on top of a virtual machine to appear on a separate device. In addition to the host operating system beneath it, this device has its operating system, libraries, and other programs.

It is used in computers for a variety of reasons. The maximum common application for desktop users is accessing software designed for a different operating system without having to switch machines. Server administrators can run various operating systems through virtualization. More importantly, virtualization allows an extensive plan to be segmented into smaller pieces, making it easier to use the server by multiple users or applications with varying requirements. It also will enable programmes inside a virtual machine to be isolated from those running in another virtual machine on the same host.

Server virtualization (OS virtualization) involves customizing a conventional operating system to enable multiple applications simultaneously on a single computer through operating system virtualization (OS virtualization). They don't conflict even if they're on the same machine [3 - 7].

The operating system behaves like numerous separate systems in a virtualized environment. Other users running other applications on the same workstation can send commands to the virtualized environment. Virtualized operating systems handle each user's requests independently.

At the operating system level, this is also known as virtualization. With operating system virtualization, applications are separated from the operating system, enabling users to enjoy application-transparent virtualization. In addition to providing granular control at the application level, OS virtualization technology allows for more flexibility and lower overhead with its improved granularity migration.

Critical applications can also be moved to another operating system instance using OS virtualization. Patches and changes to the underlying operating system are applied promptly and have minimal or no impact on application service availability. The OS virtualized environment's processes are segregated, and their interactions with the underlying OS instance are tracked.

Since hypervisors serve as a software layer, they allow one host device to support multiple virtual machines simultaneously, one of the critical components of cloud computing technology. By making cloud-based applications accessible in a virtual environment, hypervisors allow IT to maintain control over the infrastructure, processes, and sensitive data within a cloud environment.

Creative applications are becoming increasingly important in response to digital transformation and increasing consumer expectations. As a consequence, many businesses are switching to cloud computing [9]. As a result, rewriting any existing application for the cloud will take valuable IT resources and create infrastructure silos.

As part of a virtualization platform, a hypervisor also helps companies move applications quickly to the cloud. A quicker return on investment is possible by taking advantage of the cloud's many benefits, such as lower hardware costs, greater accessibility, and increased scalability [8 - 12].

Hypervisors have many Benefits

Some of the perks are as follows for hosting several virtual machines on a hypervisor:

Speed

Virtual computers using hypervisors, unlike bare-metal servers, can be constructed in seconds. It becomes considerably easier to provision resources for complex tasks.

Efficiency

Using hypervisors to run multiple virtual machines simultaneously on one physical machine tends to maximize the effectiveness of a single physical server.

Flexibility

The hypervisor separates the operating system from its underlying hardware, so the program does not rely on specific hardware devices or drivers. Operating systems and applications can run on bare-metal hypervisors regardless of the hardware type.

Portability

A hypervisor (host machine) allows running multiple operating systems on the same physical server [13 - 15]. Due to their separation from the physical computer, the virtual machines created by a hypervisor are portable.

Virtualization software enables an application to access additional machines without interruption as it requires more computing power.

CONTAINER VS HYPERVISOR

Additionally, containers and hypervisors speed up and optimize systems. Yet they each accomplish these tasks very differently, which is one of the reasons why they are different. Virtualization technology allows an operating system to operate independently of its underlying hardware and provide centralized computing, storage, and memory services [16 - 20].

There is no specific OS requirement for the container to work, as the container makes sure that the program runs. It takes a container engine to run on any platform or operating system. Containers offer unparalleled versatility since an application has everything it needs to work within a container.

Containers and hypervisors perform various functions. The only thing a container contains is an app and any services it associates with it. These lightweight and compact machines make them ideal for creating and moving applications quickly and in a flexible manner.

Type 1 Hypervisors (Bare Metal)

Hypervisors of Type 1 are installed right on the host hardware. Such hypervisors are very good for businesses because they don't need an operating system or a device driver to get to the hardware. The implementation is also built to be safe against OS-level flaws. ESXi, Microsoft Hyper-V, Xen and Oracle VM are all Type 1 hypervisors, and they all run on computers.

Type 2 Hypervisors (Hosted Hypervisor

These hypervisors run as an app on a computer with a traditional OS. If you're a developer or a person who works with security or needs apps that work on a specific computer version, you might use Type 2 hypervisors. There are two types of servers: KVM, which runs on top of an Oracle VM, and Virtual Box where there are three different programmes that run on top of Microsoft Virtual PC, but they aren't the same. The two most popular Type 2 hypervisors are QEMU and Oracle VM Virtual Box.

HYPER-V ARCHITECTURE

When using Microsoft's Hyper-V, you can run many different programmes simultaneously. A virtualization technology called Microsoft Hyper-V is used for x64 versions of Windows Server. Hyper-V runs on a hypervisor. If you want to use Hyper-V Server, you can buy a separate Hyper-V Server product. You can also add it as a role or component to Windows Server. Microsoft Hyper-V is not necessary, no matter which one of these you use. The hypervisor is the same, no matter what version of the software you have. For Microsoft Hyper-V to work, it needs a processor that can help with Virtualization on the hardware level. This allows for a smaller codebase and better performance. It is built on micro-kernelized hypervisors. This is shown in Fig. (**2**). One way to do this is to have an operating system called the "parent partition," which has features like management tools and driver downloads for the hardware.

Moreover, if Hyper-V is enabled, latency-sensitive, high-precision programmes may experience problems executing in the host. Because virtualization is enabled, the host OS, like guest operating systems, runs on top of the Hyper-V virtualization layer. However, unlike guests, the host OS has direct access to all hardware, which means that applications with particular hardware needs can function in the host OS without problems.

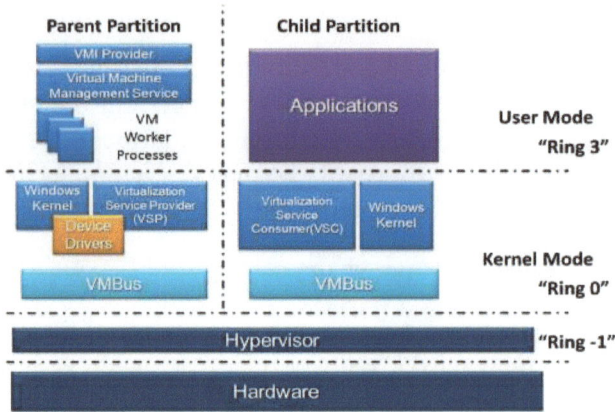

Fig. (2). Hyper-V architecture.

VMWARE ARCHITECTURE

VMware ESXi is the next era of the virtualization foundation that the company has built. It's called that because it's the next step. Because ESXi doesn't have the Linux-based Service Console, it doesn't need to be set up and run. This gives ESXi a small footprint of about 70 MB. Even though it doesn't have the Service Console, ESXi still has the same virtualization features as the earlier version of VMware ESX did. VMware ESXi needs to be used less often than before to have the same wide range of virtualization features as ESX without the Service Console. Still, the core of virtualization functionality is not found in the Service Console. You start by setting up the VMkernel. The VMkernel helps the Virtual machines (VMs) use the physical hardware by managing memory, scheduling CPUs and processing virtual switch data to get to it as shown in Fig. (**3**).

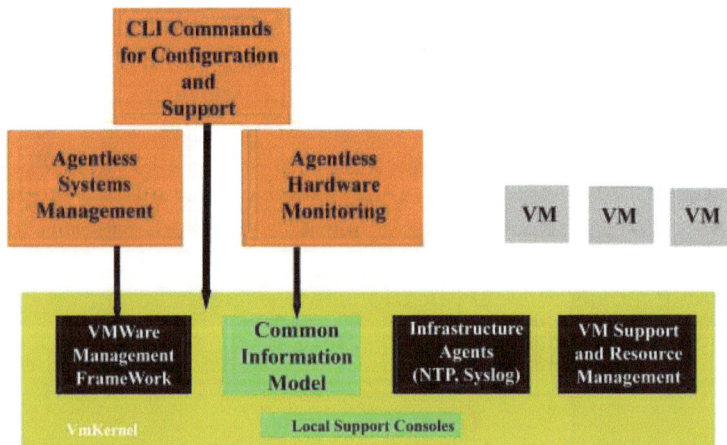

Fig. (3). VMware architecture.

VMware's virtualization package is called vSphere. vMotion, ESXi, vCenter Server, and vSphere Client are all included in VMware vSphere, which was previously known as VMware Infrastructure. Customers may run a cluster of vSphere hosts with vSAN and NSX in an Amazon data centre and run their workloads there while still managing them with their familiar VMware tools and abilities with VMware Cloud on AWS.

XEN ARCHITECTURE

Xen comprises many different parts that work together to make the virtualization environment. These three parts are called Dom0 and DomU. They can be Para-virtualized (PV) or Fully-Virtualized (FV)/Hardware Assist (HA), and they can be either (HW-Assisted) Guest A figure called Figure 4. One layer of software is called the "Xen hypervisor." This layer of software runs on the hardware below any operating system. For example, it decides which VMs get to use which CPUs and how much memory each receives. Lightweight: It can give guest domains (DomU) to the domain that has more power (DomP) (Dom0). To run Xen, the Xen hypervisor first takes control of the system. The hypervisor then loads the first guest OS, Dom0.

The following components make up the Xen virtual environment:

• Xen Hypervisor - The Xen Hypervisor constructs and controls partitions, which are segregated execution environments. Because virtual machines share a shared processing environment, the hypervisor manages their execution.

• Domain 0 - Domain 0 is a virtual machine running on the Xen Hypervisor with unique permissions to access physical I/O resources and interact with the other virtual machines on the system (Domain U: PV and HVM Guests). Domain 0 must be running in all Xen virtualization environments before any other virtual machines may get started as shown in Fig. (4).

MOTIVATION

This research looked at several different jobs with their own set of quality-o--service requirements (*e.g.*, deadline, priority and workload). Therefore, selecting a suitable virtual machine from a diverse resource set with minimal energy consumption and a task's met QoS has become a complex scheduling problem.

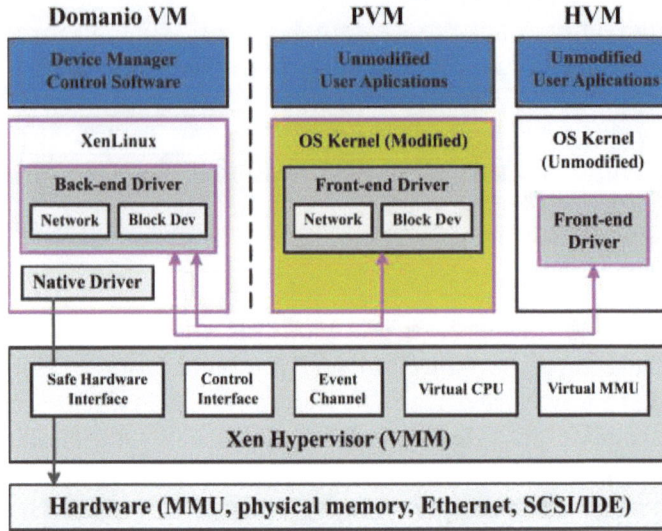

Fig. (4). Xen Architecture.

• This research provides a system architecture for processing jobs on heterogeneous computing virtual machines under tight deadlines using various components. The system comprises multiple virtual machines, and the scheduler will determine how and where positions will be assigned based on the quality-o--service criteria.

• The initial phase in the scheduling problem is the work sequence rule, and the study generates new rules to ensure the quality of services required during scheduling.

PROBLEM STATEMENT

One of the most critical technologies in cloud computing systems is Operating System virtualization. This technique allows several virtual computers to be installed and run on a single physical machine. Requests are reviewed now and then and formed into virtual machines, subsequently assigned cloud infrastructure resources [14, 15]. A pool of physical machine resources with varying capacities exists in a cloud computing environment. The virtual machine placement challenge is mapping several virtual machines to a group of physical computers. This procedure is critical for optimal resource utilization and energy consumption in cloud infrastructure. However, offering an efficient solution is not an easy task due to request and physical machine heterogeneity, multi-dimensionality of resources, and enormous scales. This mapping must be created to meet the primary requirements of a data centre, such as reducing energy usage and costs while increasing profits. (Fig. **1**).

METHOD

We did experiments in two dimensions.: Workload and Dynamic Memory. For example, we ran different hypervisors on a piece of hardware. Each hypervisor had several virtual machines running two different workloads and memory, and each hypervisor had a lot of memory. Our method gives us a chance to look into the problems we face. The response time of a VM and our experiments show that the response time and memory consumption change significantly on the same hardware with different hypervisors running on it.

EXPERIMENTAL SETUP

To comprehensively compare the hypervisor's workload and dynamic memory efficiency, we selected three different hardware platforms as our testbed. The testbed configuration is given in Table 1.

Table 1. Platform configuration.

Platform	Laptop	Cores	Memory	Storage	CPU
Dell X66 Based PC	Laptop	4	8 GP	512 SSD	Intel i5
Hp X64 Based PC	Desktop	4	8 GP	512 SSD	Intel i3
HP	Laptop	4	4 GP	512 SSD	Intel i3

EXPERIMENT WORKLOADS AND ITS CLASSIFICATION

On top of the hypervisor, we ran two different Virtual Machines with two different workloads. To figure out prime numbers, we used a program written in Java. Java, has two types namely, simple prime and complex prime. Simple Prime is an ordinary program to find whether the given number is prime or not. The second Complex Prime is written using recursion, executed for prime search; it calculates and searches for prime numbers in 10 intervals. Each interval will have 10000 ranges. The response time of each execution was calculated and noted, and taken an average of them for comparison purposes. The Dynamic memory Pressure was calculated for each virtual machine by monitoring the hypervisors console of each hypervisor.

RESULTS AND DISCUSSION

Observation 1

We have compared the response time of Hyper -V, XenServer, KVM, VMwareon various platforms, including a desktop and a laptop. First, we run two different prime searches on the platform. The simple Prime and complex prime search have given different response time values. The values are noted for each hypervisor, as shown in Fig. (**5**). An overall analysis of four different hypervisors is given in Fig. (**6**).

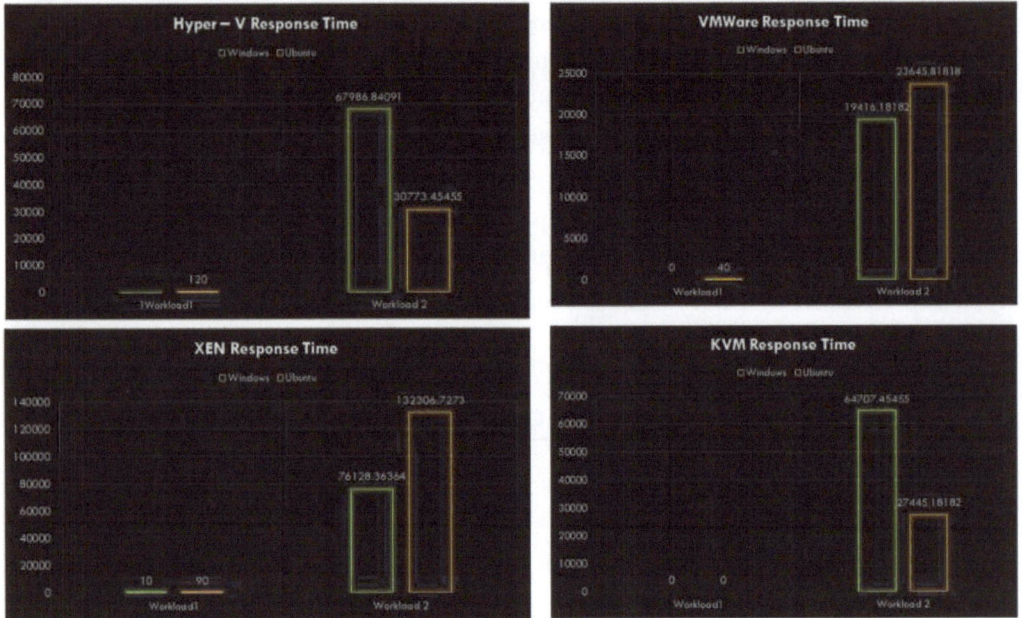

Fig. (5). Response time of hypervisors.

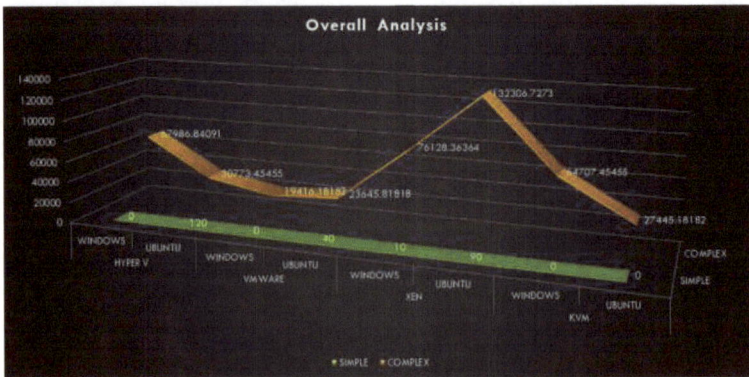

Fig. (6). Overall analysis of response tim.

Observation 2

We have compared the dynamic memory pressure of Hyper -V, Xen Server, KVM, and VMware on various platforms, including a desktop and a laptop. We ran computation-intensive in these platforms to investigate the memory pressure when two different workloads create differences in these hypervisors. In addition, we tried to make the environment look like natural multi-tenant cloud computing environment with virtual machines with varying workload levels Fig. (7).

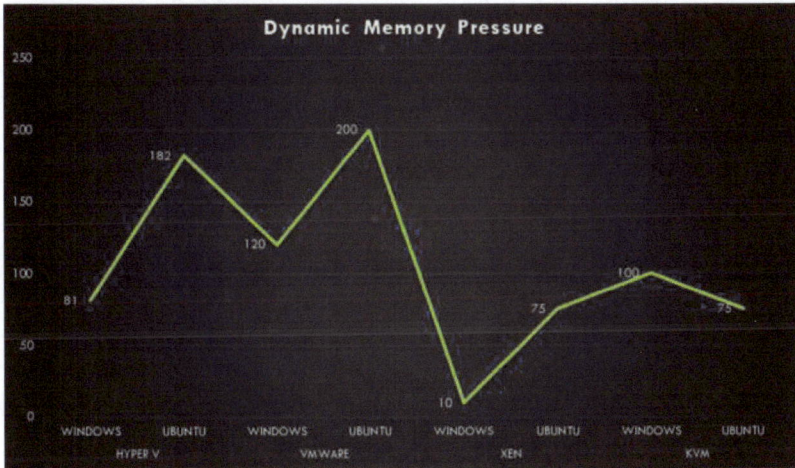

Fig. (7). Overall dynamic memory pressure of hypervisors.

OVERALL RESULT ANALYSIS

From observation one and observation two, we can come to an analysis that:

a. The hypervisor, hardware, and type of workload are all connected [4]. This caused design engineers to be careful when choosing hypervisors for virtualized infrastructure and cloud data centres.

b. Objective 1: From Table **2**, objective 1, the response time values for Windows-VM and Ubuntu-VM are less for the KVM hypervisor.

Table 2. Response time values of type 1 hypervisors.

Workload/VM	HYPER V		XEN		KVM	
	WINDOWS	UBUNTU	WINDOWS	UBUNTU	WINDOWS	UBUNTU
SIMPLE	0	120	10	90	0	0
COMPLEX	67986.84091	30773.45455	76128.36364	132306.7273	64707.45455	27445.18182

c. Objective 2: From the Table **3**, the objective 2 Dynamic Memory Values for Windows-VM and Ubuntu-VM are less for XEN hypervisor.

Table 3. Memory pressure values of hypervisors.

Memory/VM	HYPER V		VMWARE		XEN		KVM	
	WINDOWS	UBUNTU	WINDOWS	UBUNTU	WINDOWS	UBUNTU	WINDOWS	UBUNTU
Memory	81	182	120	200	10	75	100	75

d. So, the analysis shows. Type 1 Hypervisors has more advantages than Type 2 Hypervisor because Type 2 is built on the top of OS of architecture is more complicated and need to make more adjustment.

CONCLUSION

Nowadays, numerous virtualization platforms range from open-source hypervisors to commercial ones. These hypervisors can be Type1 running directly on top of the hardware or Type2 being an application in an operating system (OS). This paper provides a performance analysis of Microsoft (MS) Hyper-V, VMware ESXi and Xen based on workload and memory. Our findings show that different hypervisors have different response times and memory characteristics on the same hardware with the same workload on the same machine. Furthermore, different hypervisors have various features and work well with different workloads and workload levels. In addition, they can be used for varying levels of work in different power situations. It will be possible to test this single system hypervisor in the future in a cloud environment with even more complicated workloads, and the Type2 hypervisor will be analyzed even more. Today, there are many different virtualization platforms, from open-source hypervisors to commercial ones. These hypervisors can be either Type1 or Type2. Type1 hypervisors run on top of the hardware, while Type2 hypervisors run inside an operating system (OS).

REFERENCES

[1]   G. Aranda Osorio, C.A. Flores Valdez, M. Cruz Miranda, C. Sanchez del Real, and A. Hernandez Garay, "Effect of inclusion of cactus pear cladodes in diets for growing-finishing lambs in central Mexico", *Acta Hortic.,* no. 728, pp. 269-274, 2006.
[http://dx.doi.org/10.17660/ActaHortic.2006.728.38]

[2]   B. Dordevic, V. Timcenko, O. Pavlovic, and N. Davidovic, "Performance comparison of native host and hyper-based virtualization VirtualBox", *Int. Symp. Infoteh-jahorina, Infoteh* 17-19 March 2021, East Sarajevo, Bosnia and Herzegovina, pp. 17–19.
[http://dx.doi.org/10.1109/INFOTEH51037.2021.9400684]

[3]   H. Fayyad-Kazan, L. Perneel, and M. Timmerman, "Benchmarking the performance of microsoft hyper-V server, VMware ESXi and Xen hypervisors", *J. Emerg. Trends Comput. Inf. Sci.,* vol. 4, no. 12, pp. 922-933, 2013.

[4]   C. Jiang, Y. Wang, D. Ou, Y. Li, J. Zhang, J. Wan, B. Luo, and W. Shi, Energy efficiency comparison of hypervisors.*Sustainable Computing: Informatics and Systems* vol. 22. Elsevier, 2019, pp. 311-321.
[http://dx.doi.org/10.1016/j.suscom.2017.09.005]

[5]   G. Xiang, H. Jin, and D. Zou, "A comprehensive monitoring framework for a virtual computing

environment", *Int. Conf. Inf. Netw.* 01-03 February 2012, Bali, Indonesia pp.551-556.
[http://dx.doi.org/10.1109/ICOIN.2012.6164438]

[6]   N. Penneman, D. Kudinskas, A. Rawsthorne, B. De Sutter, and K. De Bosschere, "Formal virtualization requirements for the ARM architecture", *J. Syst. Archit.*, vol. 59, no. 3, pp. 144-154, 2013.
[http://dx.doi.org/10.1016/j.sysarc.2013.02.003]

[7]   D. Kapil, "A comparision study: XenMotion vs vmotion", *Int. J. Sci. Res.*, vol. 8, no. 2, p. 1, 2019.

[8]   T. Thein, S. Do Chi, and J.S. Park, "Availability modeling and analysis on virtualized clustering with rejuvenation", *Int. J. Comput. Sci. Netw. Security.*, vol. 8, no. 9, pp. 72-80, 2008.

[9]   Z. Inayat, A. Gani, N.B. Anuar, S. Anwar, and M.K. Khan, "Cloud-Based intrusion detection and response system: Open research issues, and solutions", *Arab. J. Sci. Eng.*, vol. 42, no. 2, pp. 399-423, 2017.
[http://dx.doi.org/10.1007/s13369-016-2400-3]

[10]   K. Ye, X. Jiang, D. Huang, J. Chen, and B. Wang, "Live migration of multiple virtual machines with rearticle-title reservation in cloud computing environments", *IEEE 4th International Conference on Cloud Computing,* 04-09 July 2011, Washington, DC, USA.
[http://dx.doi.org/10.1109/CLOUD.2011.69]

[11]   M. R. Hines, U. Deshpande, and K. Gopalan, "Post-copy live migration of virtual machines", *ACM SIGOPS Operating Systems Review,* vol. 43, no. 3, pp. 14-26, 2009.
[http://dx.doi.org/10.1145/1618525.1618528]

[12]   W. Chen, H. Lu, L. Shen, Z. Wang, N. Xiao, and D. Chen, "A novel hardware-assisted full virtualization technique", *Proc. 9th Int. Conf. Young Comput. Sci. ICYCS 2008* 18-21 November 2008,Hunan, China, pp.1292-1297.
[http://dx.doi.org/10.1109/ICYCS.2008.218]

[13]   M.M. Ahmadi, F. Khalid, and M. Shafique, "Side-Channel Attacks on RISC-V Processors: Current progress, challenges, and opportunities", *arXiv,* vol. 2106, p. 08877. Available from: http://arxiv.org/abs/2106.08877

[14]   J. Y. Hwang, "Xen on ARM: System virtualization using Xen hypervisor for ARM-based secure mobile phones", *5th IEEE Consumer Communications and Networking Conference,* 10-12 January 2008, Las Vegas, NV, USA,
[http://dx.doi.org/10.1109/ccnc08.2007.64]

[15]   J. Shuja, A. Gani, K. Bilal, A.U.R. Khan, S.A. Madani, S.U. Khan, and A.Y. Zomaya, "Survey of mobile device virtualization: Taxonomy and state of the art", *ACM Comput. Surv.*, vol. 49, no. 1, pp. 1-36, 2017.
[http://dx.doi.org/10.1145/2897164]

[16]   J.O.F. Telecommunications, "A new model for virtual machine migration in virtualized cluster server based on fuzzy decision making response", *J. Telecommun. Inf. Technol.*, vol. 1, no. 1, 2010.

[17]   B. Annane, and O. Ghazali, "Virtualization-based security techniques on mobile cloud computing: Research gaps and challenges", *Int. J. Interac. Mob. Technol.*, vol. 13, no. 4, pp. 20-32, 2019.

[18]   F. Zhang, G. Liu, X. Fu, and R. Yahyapour, "A survey on virtual machine migration: Challenges, techniques, and open issues", *IEEE Commun. Surv. Tutor.*, vol. 20, no. 2, pp. 1206-1243, 2018.
[http://dx.doi.org/10.1109/COMST.2018.2794881]

[19]   S. Deb, K. Chang, X. Yu, S.P. Sah, M. Cosic, A. Ganguly, P.P. Pande, B. Belzer, and D. Heo, "Design of an energy-efficient CMOS-compatible NoC architecture with millimeter-wave wireless interconnects", *IEEE Trans. Comput.*, vol. 62, no. 12, pp. 2382-2396, 2013.
[http://dx.doi.org/10.1109/TC.2012.224]

[20]   Avaialble   from: https://www.infoworld.com/article/2635976/xensource-versus-vmware-in-performance-comparison.html

CHAPTER 7

A Study on Load Balancing in Cloud Computing

M. Vidhya[1,*] and **R. Devi**[1]

[1] *Department of Computer Science, Vels Institute of Science, Technology and Advanced Studies (VISTAS), (Deemed to be University), Chennai, India*

Abstract: Cloud computing provides a dynamic model that provides many more services to users, as well as organizations, that can purchase based on their requirements. Cloud offers services such as storage for data, a platform for application development and testing, providing an environment to access web services, and so on. Common issues in a cloud environment are maintaining the application performance with Quality of Service (QoS) and Service Level Agreement (SLA) provided by the service providers to the organization. The major task done by the service providers is to distribute the workload among multiple servers. An effective load-balancing technique should satisfy the user requirements through efficient resource allocation in Virtual Machines. A review of various LB techniques that result in overall performance and research gaps is discussed in this paper.

Keywords: Architecture, Cloud computing, Issues, Load balancing, Metrics.

INTRODUCTION

Cloud computing provides the processing environment over the internet to clients or organizations based on client requirements at any time. Clients can utilize resources whenever they need an on-demand service. Cloud provides extendable services of distributed as well as parallel computing. Various features of the cloud are flexibility, scalability, on-demand services, *etc.*, Virtualization is the main concept implemented in the cloud concept [1]. Cloud Service Providers (CSP) provide a variety of computing environments such as SaaS (Software as a Service), PaaS (Platform as a Service), and IaaS (Infrastructure as a Service) to cloud clients. Clients can access all kinds of services on pay per use basis.

CSP offers Service Level Agreement (SLA) documents to clients for satisfactory services level [2].

---
[*] **Corresponding author M. Vidhya:** Department of Computer Science, Vels Institute of Science, Technology and Advanced Studies (VISTAS), (Deemed to be University), Chennai, India; E-mail: mvidhyavenkat11@gmail.com

S. Kannadhasan, R. Nagarajan, N. Shanmugasundaram, Jyotir Moy Chatterjee & P. Ashok (Eds.)

Four deployment models are present in clouds such as public, private, community, and hybrid clouds [3]. The public model is available for all public provided by service providers. The private cloud is exclusively utilized only by a single organization. The community cloud is used by a group of enterprises that have the same requirements and work nature. A hybrid cloud is a combination of private and public clouds that uses some strategies for application and data availability between those environments.

Major issues in Cloud Computing include efficient resource allocation, security, effective load balancing, and support level of scaling. In the cloud, the workload can be balanced properly is an important concern [4]. The main function of a task scheduler is to monitor every virtual machine and assign the task to the virtual machine with the required resources. Load balancing algorithms can be used to handle the load for the virtual machine without underload or overload. The main purpose of using load balancing algorithms is to utilize the resources effectively and complete the tasks with minimal makespan and increase throughput [5]. Tasks assigned to the virtual machines in the computing environment use either static or dynamic scheduling.

CLOUD COMPUTING ARCHITECTURE

CC architecture slightly differs from the traditional distributed and parallel system architecture. CC architecture supports high-level scalability, provides a variety of distinct services to the client, and enables the dynamic required services through virtualization. The single host system architecture is represented in Fig. (1). Architecture has 3 layers: a Hardware layer, a Virtualization layer, and an Application layer. The hardware layer is designed with virtualized hardware resources such as memory, network, and processing unit [6]. In the virtualization layer, the Hypervisor or virtual machine monitor acts as a mediator between the multiple virtual machines and guest OS. A hypervisor is a software that supports multiple platforms in a single hardware. In the application layer, the customers can use or create the application and test it under multiple platforms [7]. Each VM processes a single task at a time. Among multiple VMs, one acts as a load balancer, it allocates the task to VM if the requested resources are available to complete the task within the allocated deadline.

Multiple users can submit their tasks with different resource requirements. To handle the environment cloud service provider maintains the scheduling model that is shown in Fig. (2). All tasks are entered into the task queue. VM Manager receives the task from the task queue and checks the required resource availability for all tasks [8]. If the VMs are available, the task is assigned to the task scheduduler. Mapping among the tasks is done by the task scheduler based on resource availability. A finite number of VMs only present in every host.

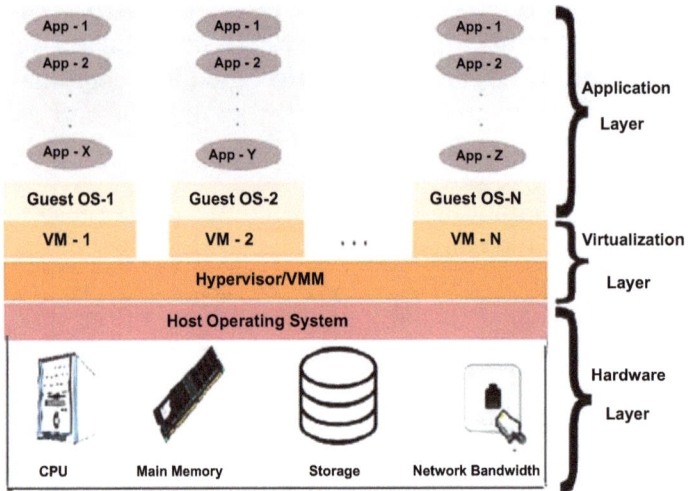

Fig. (1). Single Host Architecture in CC.

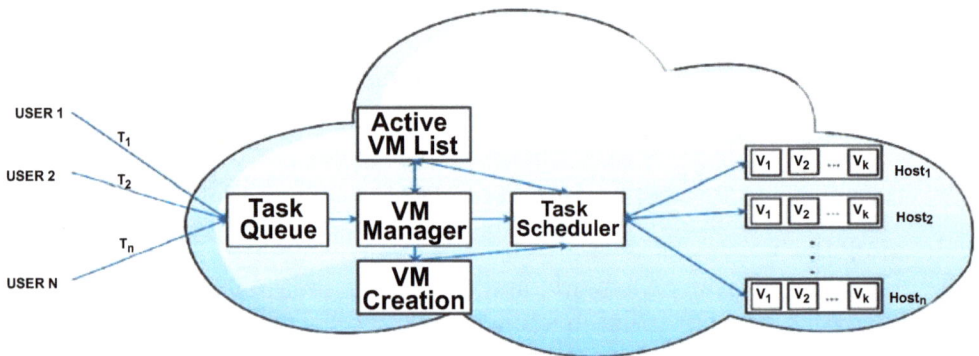

Fig. (2). Scheduling Model.

LOAD BALANCING

In a cloud environment, a method used to optimize the resources is load balancing. It handles the dynamic distribution of workload and utilization of resources effectively. We can reduce the underloaded and overloaded situation of nodes by implementing the load balancing technique. The load balancing model can be described in Fig. (**3**) [9]. Multiple client requests can be received by the Datacenter Controller (DCC), which acts as the job manager. Tasks are forwarded to the load balancer which has the load balancing algorithms to allocate the task to VMs. Virtual Machine Manager maintains a list of underloaded VMs. VM is a software installed on top of the host then we can execute the OS and application.

Tasks are received by VMM and then allocates to the underloaded VM on that physical machine [10]. To attain good performance, CSP should invoke better load-managing algorithms. If the client is not satisfied with the current cloud performance, the client may change their services to any other CSP.

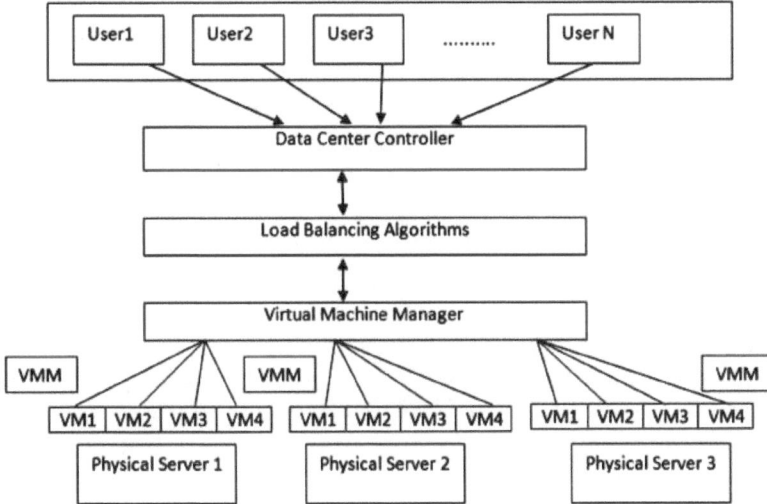

Fig. (3). Load Balancing Model.

Load balancing is a necessary technique to assure the QoS in cloud data centers. Many issues occur on the LB, among them a few issues can be specified as follows:

Graphical Distribution Node

In the cloud, the data centers are geographically distributed otherwise multiple distributed nodes are treated as a single node without considering any factors such as any delay in communication or network or gap present between the sender and the receiver node [11].

Virtual Machine Migration

In a single physical node, we can design multiple VMs. By this VM structure, the host machine can be overloaded.

Algorithm Complexity

Load balancing algorithms should be designed in a simple understandable format that should attain high efficiency.

Heterogeneousous Nodes

Requirements of different clients may vary. To satisfy all client needs, we need different kinds of nodes that provide a variety of services. Load balancing techniques are used to achieve that by enforcing the heterogeneous types of nodes.

Single Point of Failure

Among multiple nodes, one node acts as a load balancer. It assigns the tasks to other nodes [12]. If any failure occurs on the centralized controller node, the entire system functionality is down.

Load Balancer Scalability

Load balancer scalability should depend on the computing power, used network topology, and the availability of storage capacity.

METRICS FOR LOAD BALANCING IN THE CLOUD

Various metrics are considered before enforcing the load-balancing algorithms to enhance the efficiency of the system [13].

Throughput

The number of jobs finished per unit time in a VM is considered the throughput of a VM. Well-performed VM only gets high throughput.

Response Time

The total time spent for transmission, waiting, and service. Through quick response, only we will get a shorter makespan.

Makespan

The time taken to complete all the allocated processes to a VM is treated as makespan. Through optimal load balancing, we attain less makespan.

Scalability

It specifies the system's ability to adapt to the dynamically changed client requirements. The system should maintain its performance even if the workload increases. Several resources will be changed frequently in the cloud, so the system should be flexible to adapt to that.

Fault Tolerance

For any failure that happens on system components; we should have a fault-tolerance method to handle and recover the fault. Failure can occur on a single point or multipoint. CSP should maintain extra VMs as well as resources to manage the fault, and maintain the system as fault-free.

Migration Time

Time taken to transfer the job from one VM to another VM. Migration can be taken on multiple VM on the same host or multiple VM on different hosts.

Energy Consumption

Using load balancing, CSP should balance the workload across all nodes in the cloud and reduce the energy consumption of all system resources.

Carbon Emission

It calculates carbon dioxide emitted by the system resources. Load balancing sends the task to under loaded node. Through energy efficiency, we can reduce carbon emissions.

LOAD BALANCING POLICIES

In load balancing, we can invoke many policies on it.

Selection Policy

This selection policy is invoked to transfer the task from one VM to any other VM if the node is overloaded.

Location Policy

This policy sends the task to an available underloaded VM with the required resources.

Transfer Policy

This policy has shifting conditions for the tasks from one VM to another VM or remote VM.

Information Policy

This policy maintains the information regarding all the resources and decisions that provide beneficial results.

ISSUES IN LOAD BALANCING

While working in a cloud environment, we should face many critical issues to overcome. The main challenges are data loss, load balancing, and security. In load balancing, it provides a method to allocate newly arrived tasks to VM, which completes the execution with minimal time. Several algorithms are designed to attain a high throughput with effective resource utilization [14]. Some other algorithms are designed to attain less makespan, execution time, and response time. Many static, as well as dynamic algorithms, are used to allocate the task across the network. Dynamic algorithms are state-independent but comparatively more complex than static algorithms. Identifying each node state is difficult. Migration details and approaches cannot be predicted how tasks shift from one node to another. Using the following approach, we can identify how tasks are migrated from one node to another.

Data Gathering Rules

It describes what information is to be collected and when that information is to be collected and how it can be collected.

Picking Rules

Identify which task creates system overload and then moves the task from that VM to another VM.

Aggravating Rules

It defines the task balancing duration that can be implemented and managed.

Place Rules

Identify the correct VM to assign the arriving task.

Migration Rules

This function determines whether the node is suitable for migration.

SURVEY OF CURRENT LOAD BALANCING ALGORITHMS

Load balancing algorithms can be static or dynamic or a hybrid of both, which aim to attain the best performance is shown in Table **1**.

Table 1. Concepts and research gaps present in various cloud papers.

Reference Papers	Technique Implemented	Compared With	Description	Concept	Tool Used	Research Gap
Load Balancing in Cloud Computing [6]	Cloud Load Balancing Algorithm	-	This approach to load balancing in cloud computing and the aims of the algorithm are to attain an equal workload for all servers.	Improves performance by using parameters such as resource utilization, response time, and throughput.	CloudSim	It should be enhanced as cost-effective in the future.
Load Balancing in Cloud Computing Environment Using Improved WRR Algorithm [15]	Improved WRR	RR and WRR	IWRR suitable for heterogeneous environments with the same or different nature of jobs.	The method reduces response time	CloudSim	The completion time of each job can be compared in the different scheduling and load balancing algorithms.
Dynamic time quantum priority-based RR for load balancing in cloud environment [16]	Time Quantum Priority Based Round Robin	Priority Based DRR, Fair RR, MRR	Determine dynamic time quantum (TQ) for every service request.	Priority has been considered	-	Aim to implement on the real-time system or in any simulator.
C Luster - Based L Oad B Alancing a Lgorithms [17]	Priority Algorithm	RR	Achieved better resource utilization, Response time, and improved system performance.	Avoid a large number of jobs in the wait queue	CloudSim	Extend the study and consider QoS Parameters such as cost, and waiting time.
Efficient load balancing using an improved central load balancing technique [18]	Improved Central Load Balancing Technique	ACO algorithm	"Central Load Balancer" is used to balance the workload among virtual machines in the data center.	Avoid VM overloaded	Cloud Analyst	The selection of a better machine is possible without resource wastage.

(Table 1) cont.....

Reference Papers	Technique Implemented	Compared With	Description	Concept	Tool Used	Research Gap
Dynamic Task Scheduling Algorithm Improved by Load Balancing in Cloud Computing [19]	Dynamic Task Scheduling Algorithm	RR and Honey Bee	A dynamic method for task scheduling for VMs to attain load balancing and reliability	Increase reliability and reduces the makespan.	CloudSim	The proposed method reduces the waiting time. In the future, they aim to implement this in the workflows and concentrate on other factors like the cost for CSP.
Load Balance in Cloud Computing using Software Defined Networking [20]	Load Balance using Software Defined Networking	RR	HTTP server's traffic can be distributed among multiple servers instead of one server.	Enhance the availability and improve the latency	Open Stack cloud	Load balancing can be applied on any other topology or we may use any other proxy server instead of HAproxy.
A Load Balancing with Fault Tolerance Algorithm for Cloud Computing [21]	Fault Tolerance Algorithm	Priority algorithm	The method that combines a load balancing algorithm with a fault tolerance technique to enhance availability achieves a better Performance.	Reduces response time in case of failure.	Java	Implemented over multiple data centers and the replica runs on the different datacenter.
An energy efficient load balancing on cloud computing using adaptive cat swarm optimization [22]	Adaptive cat swarm optimization	Cat Swarm Optimization, GA	To overcome the optimization issues using adaptive cat swarm optimization (ACSO) algorithm.	Concentrates on migration cost and storage utilization.	CloudSim	Implement this methodology in real-time.

(Table 1) cont.....

Reference Papers	Technique Implemented	Compared With	Description	Concept	Tool Used	Research Gap
A Load Balancing Algorithm for the Data Centres to Optimize Cloud Computing Applications [23]	Data Centre Optimization	Existing Dynamic LBA algorithm.	Improves Load Balancing Quality by means of the priority of VMs, and resource allocation.	It attains good performance in terms of less Execution time and Makespan.	CloudSim	Enhance cloud-based application performance by considering more SLA parameters.
Cloud Load Balancing Algorithm [24]	Improved Central Load Balancing	ACO	"Central Load Balancer" is to balance the burden among virtual products with the reasoning data center.	Concentrates to support scalability and avoid resource wastage.	Cloud Analyst	Improvise to lead to cost-effective results.
Dynamic load balancing algorithm to minimize the makespan time and utilize the resources effectively in cloud environment [25]	Dynamic load balancing algorithm	FCFS, SJF, Min-Min	A dynamic load balancing algorithm that utilizes resources properly with minimal makespan.	Concentrates on resource utilization	CloudSim	It should consider task priority, task deadline, and other quality of service (QoS) parameters
An efficient load balancing system using adaptive dragonfly algorithm in cloud computing [26]	Adaptive dragonfly algorithm	Firefly and Dragonfly algorithm	A combination of the dragonfly algorithm and firefly algorithm to attain better performance and multi-objective function is developed.	Concentrates completion time, processing costs, and load.	cloud Sim	Implement a real-time system.

CONCLUSION

In the cloud, load balancing on VMs is a primary issue identified by most researchers. This paper identifies the problems that occur with load balancing. System architecture for the host is described. Various measures that identify the system's performance are illustrated. We specify the policies invoked on load balancing in the cloud. In a separate section, we illustrate the techniques used to implement the new algorithm. The newly created algorithm is compared with existing algorithms, with respect to the purpose of their design, what they attain, and what tools are used to implement and specify the research gap.

REFERENCES

[1]   D. A. Shafiq, N. Z. Jhanjhi, and A. Abdullah, "Load balancing techniques in cloud computing environment: A review", *J. King Saud Univ. - Comput. Inf. Sci.,* vol. 34, no. 7, pp. 3910-3933, 2021. [http://dx.doi.org/10.1016/j.jksuci.2021.02.007]

[2]   S. Shiju George, and R. Suji Pramila, "A review of different techniques in cloud computing", *Mater. Today Proc.,* vol. 46, pp. 8002-8008, 2021. [http://dx.doi.org/10.1016/j.matpr.2021.02.748]

[3]   U. Dubey, and L.S. Songare, "Analysis of load balancing in cloud computing", *Int. J. Sci. Technol. Res.,* vol. 8, no. 12, pp. 3912-3914, 2019.

[4]   M. Yadav, and J.S. Prasad, "A review on load balancing algorithms in cloud computing environment", *Int. J. Comput. Sci. Eng.,* vol. 6, no. 8, pp. 771-778, 2018. [http://dx.doi.org/10.26438/ijcse/v6i8.771778]

[5]   K. Balaji, P. S. Kiran, and M. S. Kumar, "Load balancing in cloud computing: Issues and challenges", *2nd International Conference on Contemporary Computing and Informatics (IC3I),* 14-17 December, Greater Noida, India, 2016, pp. 120-125.

[6]   S.K. Mishra, B. Sahoo, and P.P. Parida, "Load balancing in cloud computing: A big picture", *J. King Saud Univ. - Comput. Inf.,* vol. 32, no. 2, pp. 149-158, 2020. [http://dx.doi.org/10.1016/j.jksuci.2018.01.003]

[7]   A.A.A. Alkhatib, A. Alsabbagh, R. Maraqa, and S. Alzubi, "Load balancing techniques in cloud computing: Extensive review", *Adv. sci. technol. eng. syst. j.,* vol. 6, no. 2, pp. 860-870, 2021. [http://dx.doi.org/10.25046/aj060299]

[8]   S. Swarnakar, N. Kumar, A. Kumar, and C. Banerjee, "Modified genetic based algorithm for load balancing in cloud computing", *Int. Conf. Converg. Eng. ICCE,* 05-06 September 2020, Kolkata, India, pp. 255–259. [http://dx.doi.org/10.1109/ICCE50343.2020.9290563]

[9]   B. Liu, J. Chang, L. Xiao, G. Qin, B. Wei, and Z. Huo, "DDLB: A dynamic and distributed load balancing strategy", *Int. Conf. High Perform. Comput. Commun. 17th IEEE Int. Conf. Smart City 5th IEEE Int. Conf. Data Sci. Syst,* 10-12 August 2019, Zhangjiajie, China, pp. 1928–1936. [http://dx.doi.org/10.1109/HPCC/SmartCity/DSS.2019.00266]

[10]   L. Shen, J. Li, Y. Wu, Z. Tang, and Y. Wang, "Optimization of artificial bee colony algorithm based load balancing in smart grid cloud", *IEEE PES Innov. Smart Grid Technol. Asia, ISGT,* 21-24 May 2019,Chengdu, China, pp. 1131–1134. [http://dx.doi.org/10.1109/ISGT-Asia.2019.8881232]

[11]   L. Tang, "Load balancing optimization in cloud computing based on task scheduling", *2018 Int. Conf. Virtual Real. Intell. Syst. ICVRIS 2018,* 10-11 August 2018,Hunan, China pp.116-120. [http://dx.doi.org/10.1109/ICVRIS.2018.00036]

[12]   D. Ramesh, and S. Dey, "SCLBA-CC: Slot based carton load balancing approach for cloud environment", *Proc. 2018 Int. Conf. Curr. Trends Towar. Converging Technol. ICCTCT 2018,* 01-03 March 2018, Coimbatore, India, pp.1-5. [http://dx.doi.org/10.1109/ICCTCT.2018.8550841]

[13]   Sreelakshmi, and S. Sindhu, "Multi-Objective PSO based task scheduling-a load balancing approach in cloud", *Proc. 1st Int. Conf. Innov. Inf. Commun. Technol. ICIICT 2019,* 25-26 April 2019 , Chennai,

India pp.1-5.
[http://dx.doi.org/10.1109/ICIICT1.2019.8741463]

[14] D. Malhotra, "LD _ ASG : Load balancing algorithm in cloud computing", *Fifth Int. Conf. Parallel, Distrib. Grid Comput,*, pp. 387-392, 2018.

[15] D.C. Devi, and V.R. Uthariaraj, "Load balancing in cloud computing environment using improved weighted round robin algorithm for nonpreemptive dependent tasks", *Sci. World. J.,* vol. 2016, pp. 1-14, 2016.
[http://dx.doi.org/10.1155/2016/3896065] [PMID: 26955656]

[16] S. Ghosh, and C. Banerjee, "Dynamic time quantum priority based round robin for load balancing in cloud environment", *IEEE Int. Conf. Res. Comput. Intell. Commun. Networks, Icrcicn,* 22-23 November 2018,Kolkata, India, pp. 33–37.
[http://dx.doi.org/10.1109/ICRCICN.2018.8718694]

[17] R.U. Payli, K. Erciyes, and O. Dagdeviren, "Cluster - based l oad b alancing a lgorithms", *Int. J. Comp. Networ. Commun.,* vol. 3, no. 5, pp. 253-269, 2011.

[18] S. Kaur, and T. Sharma, "Efficient load balancing using improved central load balancing technique", *Proc. 2nd Int. Conf. Inven. Syst. Control. ICISC,* 19-20 January 2018, Coimbatore, India, pp. 1–5.
[http://dx.doi.org/10.1109/ICISC.2018.8398857]

[19] F. Ebadifard, S.M. Babamir, and S. Barani, "A dynamic task scheduling algorithm improved by load balancing in cloud computing", *Conf. Web Res. ICWR 2020,* 22-23 April 2020, Tehran, Iran, pp. 177-183.
[http://dx.doi.org/10.1109/ICWR49608.2020.9122287]

[20] Y.A.H. Omer, M.A. Mohammedel-Amin, and A.B.A. Mustafa, "Load balance in cloud computing using software defined networking", *Proc. 2020 Int. Conf. Comput. Control. Electr. Electron. Eng. ICCCEEE 2020,* 26 February 2021 - 01 March 2021,Khartoum, Sudan.
[http://dx.doi.org/10.1109/ICCCEEE49695.2021.9429607]

[21] T. Mohmmed, and N. Abdalrahman, "A load balancing with fault tolerance algorithm for cloud computing", *Proc. 2020 Int. Conf. Comput. Control. Electr. Electron. Eng. ICCCEEE 2020,* 26 February 2021 - 01 March 2021, Khartoum, Sudan,
[http://dx.doi.org/10.1109/ICCCEEE49695.2021.9429597]

[22] K. Balaji, P. Sai Kiran, and M. Sunil Kumar, "An energy efficient load balancing on cloud computing using adaptive cat swarm optimization", *Mater. Today Proc.,* 2021.
[http://dx.doi.org/10.1016/j.matpr.2020.11.106]

[23] D.A. Shafiq, N.Z. Jhanjhi, A. Abdullah, and M.A. Alzain, "A load balancing algorithm for the data centres to optimize cloud computing applications", *IEEE Access,* vol. 9, pp. 41731-41744, 2021.
[http://dx.doi.org/10.1109/ACCESS.2021.3065308]

[24] H. Rai, S.K. Ojha, and A. Nazarov, "Cloud load balancing algorithm", *Proc. - IEEE 2020 2nd Int. Conf. Adv. Comput. Commun. Control Networking, ICACCCN,* pp. 861-865, 2020.
[http://dx.doi.org/10.1109/ICACCCN51052.2020.9362810]

[25] M. Kumar, and S.C. Sharma, "Dynamic load balancing algorithm to minimize the makespan time and utilize the resources effectively in cloud environment", *Int. J. Comput. Appl.,* vol. 42, no. 1, pp. 108-117, 2020.
[http://dx.doi.org/10.1080/1206212X.2017.1404823]

[26] P. Neelima, and A.R.M. Reddy, "An efficient load balancing system using adaptive dragonfly algorithm in cloud computing", *Cluster Comput.,* vol. 23, no. 4, pp. 2891-2899, 2020.
[http://dx.doi.org/10.1007/s10586-020-03054-w]

CHAPTER 8

A Survey on Facial and Fingerprint Based Voting System Using Deep Learning Techniques

V. Jeevitha[1,*] and **J. Jebathangam**[2]

[1] *Department of Computer Science, Vels Institute of Science, Technology and Advanced Studies (VISTAS), Chennai, India*

[2] *Department of Information Technology, Vels Institute of Science, Technology and Advanced Studies (VISTAS), Chennai, India*

Abstract: The current electronic voting system can be hacked easily. There are a lot of methods adopted to avoid malpractice. This research provides secured voting and avoids human intervention that results in smooth and secure conduction of elections. This research adopts biometric fingerprint recognition and face recognition of the voter for authentication. In an electronic voting system, the first step in the verification process can be easily achieved with the voter fingerprint data available in this database. The second step of verification involves the face recognition of the voter by the data already present in the database. If two-phase verification is done, the voter can proceed with the voting process and present his/her vote. Then the vote will be encrypted. This prevents fake votes and ensures perfect polling without any corruption. We have created a fingerprint-based voting system where the user does not have to take hisher ID with his/her necessary information. If the details match the previously stored information of registered fingerprints, a person is allowed to cast his or her vote. If not, a warning message will be displayed and the person is excluded from voting. In an election counting stage, the admin will decrypt and count the votes.

Keywords: Deep learning algorithm, E-Voting, Encryption and decryption, Fingerprint recognition, Facial recognition.

INTRODUCTION

Biometrics is the science and technology of measuring and analyzing the natural data. The biometric field was developed and has since expanded into a variety of physical diagnoses. These diagnostic ideas have led to the creation of fingerprints that serve to rapidly recognize people and give them access rights. The essential point of these machines is to scan the data of individual fingerprints and facial recognition and compare them with the data of other fingerprints and facial recog-

---

[*] **Corresponding author V. Jeevitha:** Department of Computer Science, Vels Institute of Science, Technology and Advanced Studies (VISTAS), Chennai, India; E-mail: jeevithav95@gmail.com

S. Kannadhasan, R. Nagarajan, N. Shanmugasundaram, Jyotir Moy Chatterjee & P. Ashok (Eds.)

nition. In this research, we have used fingerprints and facial recognition to identify voters or to prove authenticity. Since the impression of each person's thumb is different and each person's facial pattern is different, it helps to minimize errors. A database containing fingerprints and facial recognition of all voters is required. Fake votes and repeated votes are checked in this system for an accurate coding system. In addition, elections will no longer be tedious and expensive [1].

A number of problems are associated with this approach. If sick voters are unable to reach the polling station, and without a station, they cannot approach the polling station, and bad weather cannot allow a visit to the polling station, transport pool, and much more.

LITERATURE REVIEW

Smart Voting

The Aadhar website allows us to get information about individuals above the age of 18, or we may register using the voter registration form. Voters in the primary round will get an ID, a password, and a registration email address before casting their ballots. After the voter has been verified using fingerprint data in the next step, they will be given the go-ahead to cast their ballot. The voting ID will be erased after submission and there will be no additional opportunity to vote. The voter's Aadhar information will be locked in order to trace them for future access. Accordingly, the number will be rearranged [2].

Paper-based Voting

An unfilled ballot is given to each voter, who then uses a pen or marker to indicate whether they wish to support a certain candidate. Manual voting takes a lot of time, but it is simple to cast a ballot by hand, and ballot papers may be preserved for confirmation; this is still a common method of voting [3].

Handle Voting Machine

Each racquet on this unique machine is paired with a partner. Voters press a button to support the candidate. These voting devices could compute automated voting. Voters must be educated in order to avoid the border becoming useless due to its poor usability.

Direct Electronic Voting System

DRE interacts with the touchscreen; touch screen, or buttons of the voting machine to cast a ballot. On the voting records, some of them are dozing off, and the vote-counting process is swift. However, the accuracy of another DRE that does not maintain voting records is in dispute.

Punched Cards

The voter pokes a hole in an empty balloon with a metallic gap hit. It can automatically count the votes, but if there is a low voter turnout, the result can be unreliable.

Visible Voting System

This system selects the darkest mark in each vote and totals the results after each voter completes a round with their choice on an empty ballot. Vote counting on this kind of device happens quite swiftly. If the voter completes the round, it will lead to a visual check error [4].

Voting Machine Uses Fingerprint Identification

This recognition makes use of a sensor and records the information in a file. A tiny microcontroller port will be used to transmit the data to the network app once these biometric photos have been read. The input picture will be compared to an image that has already been entered into the database, or the server will provide a message, which will be shown on an LCD and serve to verify the voter's identity. If not, it exhibits LCD-like characteristics and its drawbacks [5].

Wyndham, Chen, & Das, 2016

The authors describe yet another electronic voting system that relies on cryptographic techniques and Blockchain development to provide clear results while maintaining voter protection. According to the authors, in a rare instance, the vote is recorded and notification of the vote is conveyed to the public.

Blockchain forms a test station for electronics in their frame. The co-candidate then cancels every vote, delivers the number closer to evidence that the outcomes are as expected, and allows the voter to display their vote. The idea enhances the safety and authority of mandatory voting, voter registration, and access to preliminary results, unfettered votes, and dropping vote totals. To document that a vote was cast, we utilise the square anchor. Additionally, we use native cryptographic systems to guarantee the accurate and secret counting of votes. To demonstrate to voters that their vote has been cast, the blockchain's Merkle Tree is in operation. We eliminate the possibility of voting under pressure by employing virtual polling booths and keeping receipts secret. We cannot register when the time results are received and the time the choice is made since Hawk has the notion of combining statistics till the period previously specified. An element of a smart contract is voting voters.

Shendage & Bhaskar, 2017

They note in their study book that India is a sizable democracy where elections and voting are necessary for any socially oriented society. Due to its accessibility and constant publicity, the Electronic Voting Machine (EVM) is a highly popular voting method in India. However, this voting technology is entirely inefficient and unreliable. As a result, we present the Novel Electronic Voting System, which uses biometric identification and the broadcast servers to provide the highest level of voting process security. Two portions make up the full voting framework: one is for the voting machine, and the other is for the server framework. The brain of the voting system is the Raspberry Pi 3 display B. The voting process may be managed by this little host computer [6 - 10].

The transfer server infrastructure and biometric authentication were both taken into consideration while building the voting machine. Therefore, we utilise the biometric website for Aadhar cards that the Unique Identification Authority of India (UIDAI) currently makes available to all Indians for the purpose of verification. The administration website's pre-installed biometric data (thumbnails) should be used to verify voters. The accurate voter verification occurs before voting when there are a total of six incoming votes [11]. Therefore, the framework completely ensures vote correctness and eliminates the possibility of voting-related negligence. A significant advantage of this system is that it will eventually combine the correctness of the vote cast and minimise the time required for lengthy flying divisions to make a decision, depending on the distributed server route. Thus, it establishes a trustworthy, adaptable framework that will be challenging for everyone to take into account in terms of their abilities and age, while also preventing illegitimate voting that eventually adds to the correctness of the voting framework.

Najam, Shaikh, & Naqvi, 2018

It claims that in comparison to frameworks relying on a single visible piece of information, the suggested framework employs two strategies to persuade voters to produce superior outcomes. Voters are given polarising information through distinct marker and face-based techniques. Instead of using a single parameter IDtechnology, voter-to-voter testing throughout the decision-making process offers preferable accuracy.

In order to establish a biometric format and emphasise the output throughout the voting process, the face recognition framework employs the Viola-Jones figure close to the Haar rectangle, integrating visual detection approach, and extracting critical features. In order to validate personality using GPCA (Generalised Principle Component Analysis) and K-NN (K-Nearest Neighbour), fell machine-based separators are utilised for visual acuity. Comparing excellently categorised Eigen-vectors with a biometric format pre-programmed on the decision-control panel website yields intelligent results. The suggested framework's findings demonstrate that it performs better than any other frameworks, including those that employ various class dividers or different plans, such as those based on a facial mark or different [12]. Because the suggested frame has a 91% accuracy under visible light with authorised face, it will be highly helpful for routine usage.

METHODOLOGY

The suggested idea is a web-based application, therefore some of the fundamental web-based technologies, such as website design and image processing structures that prove the necessity for software, are included. The results of a face-to-face examination and fingerprinting will determine who is eligible to vote. On the day of election, the voter may access that website.

The Authorised Selection object's fingerprint and face recognition are stored on the server. With the IP address they supply, voters may access a website. Any device utilising a website, such as a personal computer, laptop, or mobile camera, will be used to capture the voter's face and fingerprints once they open the website and click the voting button [13]. The server will accept the captured picture. The server scans every picture on this website side by side looking for similarities with those that have been registered. The voter is already registered and recognised by the electoral commission, thus he or she is permitted to vote if identical facial recognition and fingerprint are discovered. On the server, where the fingerprints are kept, they are identical.

We can identify the fingerprints of the correct voter after comparing the two images. Analysing the input against the website-stored picture yields matching fingerprints. Similar images will be shown with their ID and on the voting website, which allows voters to choose any supporting party from the voters' list. By selecting the desired group, all other choices will be disabled and the user's decision cannot be modified. The servers receive and handle votes cast by authorised voters. A supporting party's vote total will be considered. A voter doesn't have to wait a few days to learn the results of this method, and even the vote counting is quite straightforward [14].

EVM VOTING PROCESS

The way people vote on electronic voting machines is kept the same as it is on paper-based voting systems. The pre-printed ballot papers that each voter gets under the ballot paper system speak for all the candidates who have received their mark, which is the sole distinction between the EVM and the ballot paper system. When using an electronic voting system, a voter must place their mark in front of the name of the candidate they are supporting. A button must be pressed by the voter in front of their preferred candidate.

In order to keep in mind the illiterate population of India who may find it challenging to vote using an electronic voting machine if the voting procedure becomes tough, the design of the electronic voting system and the voting system were maintained the same [15]. The following categories may be used to explain how an electronic voting system works:

The control unit, voting unit, and ballot paper verification technique are the three main components of the Indian voting machine that are utilised to complete the election process. The voting officer or the presiding officer is the control unit. A control unit is located in a different voting booth, and a 5 M cable connects the voting unit with the voter turnout check unit.

• In front of those blue buttons that must be pressed in order to cast a ballot for a candidate, there is a voting unit with the candidate's name and the party or person's emblem.

• Prior to having their name added to the voters' list, voters must first get a voter ID card or other forms of identification from the Indian government. The voter then passes on to the next electoral officer, who will mark their finger with silver nitrate ink to prevent them from changing their vote afterward.

• By pushing the voting button on the control unit, the presiding officer has now turned on the voting machine. A voting unit is opened by a beep for the purpose of voting.

• Now that voting has begun, a voter enters the polling place, hits a button next to the name of the candidate they are supporting, and the voting unit closes to prevent them from casting the second ballot.

• The name of the individual who voted for the voter is written on the voter ballot paper's trail audit unit after the vote has been cast.

• Once the last voter has finished voting, the presiding officer presses the "Closing button," and then the electronic voting machine will not record any of the votes cast. The voter confirms it, and if satisfied, he or she will proceed; otherwise, the voter will need to report to the chairperson about the malfunction of the voting machine. The administrator then disconnects and saves each voting unit, VVPAT unit, and control unit individually.

• A voter record account will then be given to each voting agent by the presiding officer.

NONE OF THE ABOVE (NOTA)

In 2009, the Indian Supreme Court ordered the Indian Electoral Commission to provide voters the option of selecting "none of the above" while casting their ballots. The government first objected, but on September 27, 2013, the Supreme Court of India declared that the right to register and do none of the aforementioned in order to vote in the election must be exercised. Accordingly a button was given on a NOTA-style electronic voting system by the Indian electoral commission [16]. The Supreme Court made this decision in an effort to enhance voter turnout. Votes marked as NOTA were counted but were declared illegitimate, according to the Electoral Commission, therefore they had no effect on the election's result. The holding percentage of NOTA in the 2014 presidential election was 1% of the total vote. The National Institute of Design Ahmadabad revealed a NOTA logo on September 18, 2015, which is a paper bag with a black cross on the opposite side.

The NOTA button first debuted in the 2013 General Assembly elections, which were conducted in Delhi, the Union Territory, Mizoram, Rajasthan, and Madhya Pradesh [17]. The NOTA vote is resurrected in these provinces and cities under the authority of the 2016 Provincial, Provincial, Kerala, Pondicherry, and Provincial Council elections being 6%. Fig. (**1**) shows that NOTA received 1.1% of the votes cast during the 2014 Lok Sabha elections.

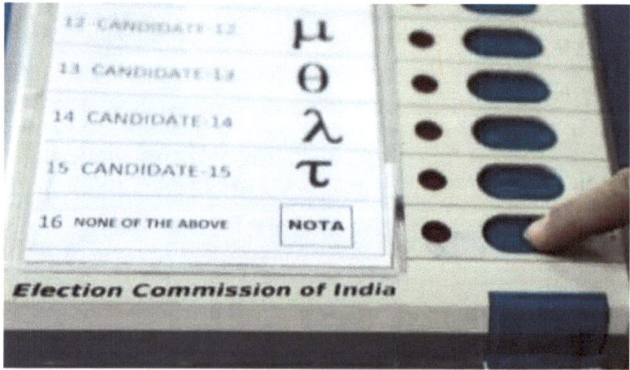

Fig. (1). NOTA.

Advantages of Electronic Voting Machine

• The option is distinct. They have a significant influence on voters' lives and views, political involvement, the governing structure, and the future of nations. Elections are obviously important to our society, thus election systems not only need to function but also for voters to think they do. The following advantages come from choosing EVM:

• Easy to Operate: India's electronic voting machines were designed with the country's almost 63.82% literacy rate in mind. These devices have a very straightforward design and are simple to use. By pressing the button in front of the candidate's emblem, those who cannot read or write may also use this to cast their ballots.

• The electronic voting machines used in India are referred to as independent systems since there is no mechanism for them to connect with other systems outside of their own. Therefore, there is no chance of hacking or remote access to EVM. A control unit and a VVPAT unit are both components of an EVM. These devices are all cable-connected and lack any additional holes for connecting to other devices.

• Eliminates Invalid Voting: In the ballot paper method, if a voter is unable to correctly mark the water criteria that must be written before the nominee's name, their vote is deemed invalid. To ensure that there is no possibility of invalid voting while using an EVM, pressure buttons are supplied, and the user must push a button.

• Transparent Auditable Secure and Reliable: India's electronic voting machine uses the VVPAT method to ensure transparency, and the voter receives information through a printed receipt known as a paper audit trail. The technology has been improved in accuracy and security to prevent remote access. This computer-generated alternative provides for a readable procedure, which implies that people may trust the outcomes.

• Increasing voter participation: The voting machine increases voter confidence, which raises the overall number of voters. Additionally, it shortens each voter's turn at the polls so that more people may vote at the same time.

• Once-customizable chip: The electronic voting system's chip is organised without enabling anybody to make ongoing changes at the same time that security measures are well-maintained in the electronic voting machine throughout hardware development [18].

• Effective key-compression encoding: EVMs employ adaptable key-compression encoding to allow them to store correct data. Before sending the data to the control unit, the voting unit creates encrypted data when a voter presses a key to cast their ballot [19]. Additionally, it has options for pushing each button in real time.

• Real Time Clock: On a fixed voting system, a real-time clock is used to time stamp each vote cast. The auditor can confirm each voter's voting time thanks to this function of the voting machine.

• A voting machine that uses electronic technology produces results quickly. India is a big nation with millions of voters, therefore it used to take days to publish the results of the ballot paper method. However, utilising an electronic voting machine, the results may be announced in a matter of hours.

Disadvantages of Electronic Voting Machine

In addition to its many benefits, the electronic voting machine has several drawbacks, such as issues with machine handling, tampering with votes while in transit, *etc.* Candidates have often suspected some kind of EVM involvement. The following are some false beliefs people have concerning computerised voting machines:

Document Recording

The absence of a site where, in the case of a booth recording, the vote is safeguarded elsewhere or on a secure worldwide server, is another significant drawback of the electronic voting system.

Avoiding Dependence on Voters

The electronic voting system used in India does not accommodate voters who are blind or visually impaired because it lacks a centre that would allow such voters to cast their own ballots in the Visually Impaired section. For the purpose of casting their votes, they depended on the attendees.

High Humidity

Using an electronic voting system in India's elite wetlands or during periods of heavy rain is not advised. EVMs should not be used in such regions since excessive humidity often causes electrical equipment to malfunction.

Participation of External Manufacturers

These pre-programmed processors are integrated in India by Bharat Electronic Limited and the Electronic Corporation of India [20]. The source code or electronic voting software was generated outside of India. The software code is in them because foreign manufacturers are involved, and they may use it to affect the election's result.

False Voting

Indian voting machines are incapable of verifying the legitimacy of a voter. Even if the voter list is not updated often, this might lead to a fraudulent vote in which a phoney voter casts a ballot in place of a legitimate one.

False Display Units

An electronic voting machine (EVM) may have a fake display unit installed that will display a fictitious result. This procedure does not need an internet criminal to compromise software; instead, owners of an EVM may simply do it by swapping out the display unit. It is a fake display unit, much like the genuine one.

Powerful Software

There is a chance to stop the voting process in progress, so you don't need any software that a computer or hacker couldn't use to manipulate the results.

Results Confirmation

Voting machines are configured in a manner that makes it impossible to check the electronic voting record by hitting a key while the votes are being counted to reflect the candidate's number of votes cast.

Postal voting is a difficulty since electronic voting machines used in elections cannot handle them. Votes are essentially transmitted by mail using postal systems. The returning officer must count the real ballots before adding them to the voter's choice. The voting procedure for those who are not present at the polling places is still in the process of printing ballots, and secondly, it takes time for these votes to be counted.

DEEP LEARNING TECHNIQUES

It employs ANNs to carry out intricate computations on a significant quantity of data. It is a kind of machine learning that mimics the structure and function of the human brain. There are several deep learning models that can accurately and successfully solve the most challenging issues in the human brain.

Recursive Neural Network

• It is called a bottomless tree-like structure. When they want to parse an entire sentence we use an RNN.

• Tree-like topology allows branch relations and hierarchical structure.

• Recursive neural networks are hierarchical kinds of networks with no time aspect to the input series but the input has to be processed hierarchically in a tree fashion.

Convolutional Neural Network

• It is an excellent tool and one of the most highly developed achievements in deep learning.

• CNNs got too much attention and focus from all major business players because of the hype of AI.

• The human brain receives any image in fractions of seconds without much effort but computer imagery is just a series of numbers.

FACIAL AND FINGERPRINT RECOGNITION

Fingerprint Recognition

It alludes to a computerised technique for comparing the fingerprints of two people. The most common biometrics are fingerprints, which are used to identify persons and ensure their individuality. An object's pattern is examined. Finger authentication may be done in many different ways. A few are copying the traditional method of enforcing the rule of matching; some do so by using the same suitable minutiae devices, while others employ even more varied methods, such as multi-line and ultrasonic devices. There are several other fingerprints that may be used for supplementary biometrics. Touch-sensitive voting machines that function in a secure environment and where you can provide users with the necessary instructions and information may benefit from fingerprint authentication. Due to the relatively low cost, compact size, and simplicity of the fingerprint verification technique to be utilised, which is shown in Fig. (**2**), it is not unexpected that the final submission area seems to depend significantly on fingerprints.

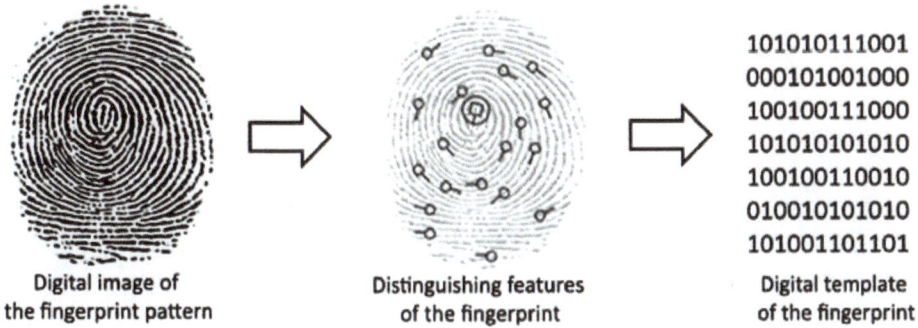

| Digital image of the fingerprint pattern | Distinguishing features of the fingerprint | Digital template of the fingerprint |

Fig. (2). Fingerprint recognition.

Face Recognition

This strategy is based on the evolutionary hypothesis, which describes the deterioration of face images. Create a small collection of features, including "Eigenfaces" photos. Fig. (**3**) depicts the major components of the face-to-face training of the training set as the Eigen Data Flow diagram of the face-based face algorithm. The Eigen facial approach is one of the most efficient and straightforward techniques for face identification. The number of photographers serves as a gauge for Eigen's distance, based on the expression on his face. If that range is lower than the indicated value, a sharp face or an unknown face is present. Two photo blocks are shown in the figure above; the first is used to train the image set, and the second is used to establish a photo clip.

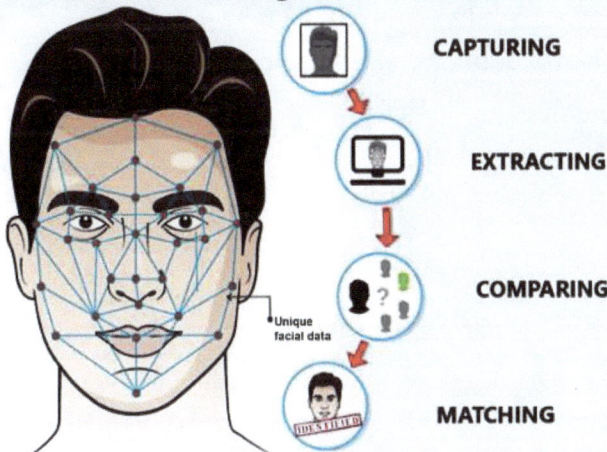

Fig. (3). Facial recognition.

CONCLUSION

Face recognition-based voting is the main component of the electronic voting-based programme. Any party size and vote distribution system in the nation may use the suggested method. In reality, LBP offers secure face authentication when used with a camera. The suggested approach is made for web-enabled smartphones and has a number of benefits over traditional voting procedures, including making it simple to vote for performers and providing accessibility for persons with disabilities, the ill, and road warriors. To solve issues with accuracy, server deployment, and system security, a number of modifications are still required.

REFERENCES

[1]   S. Jehovah, M.E. JirehArputhamoni, and A. GnanaSaravanan, "Online smart voting system using biometrics based facial and fingerprint detection on image processing and CNN", *Third International conference on Intelligent communication technologies and Virtual Mobile networks,* 04-06 February 2021,Tirunelveli, India.

[2]   A. Chowdhury, S. Kirchgasser, A. Uhl, and A. Ross, "CNN automatically learn the significance of minutiae points for fingerprint matching", *IEEE Conference* 01-05 March 2020,Snowmass, CO, USA. [http://dx.doi.org/10.1109/WACV45572.2020.9093301]

[3]   S. Agarwal, "AfreenHaider, Biometrics based secured remote electronic voting system", *IEEE Conference,* 2020 23-24 July 2020,Chennai, India

[4]   M.A. Alim, M.M. Baig, S. Mehboob, and I. Naseem, "Method for secure electronic voting system: Face recognition based approach", *Proceedings of the SPIE,* vol. 10443, 2020.

[5]   K.P. Chandra, and R. AtlaIndu, "Smart voting system using facial detection", *4th International Conference on Advances in Computing, Communication Control and Networking (ICAC3N),* 16-17 December, Noida, India, 2022, pp. 909-913.

[6]   Y.K. Al Takahashi, "Fingerprint features extraction by combining texture minutiae, and frequency spectrum using multi -task CNN", *IEEE International Joint Conference on Biometrics (IJCB),* 28 September 2020 - 01 October 2020, Houston, TX, USA, 2020, pp. 1-8. [http://dx.doi.org/10.1109/IJCB48548.2020.9304861]

[7]   T. Ayushi, and G. Neetesh, "Low resolution fingerprint image verification using CNN Filter and LSTM Classifier", *Int. J. Rec. Technol. Eng.,* vol. 8, no. 5, 2020.

[8]   Geol. Ishank, and N.B. Puhan, "Deep convolution neural network for double-identity fingerprint detection", *IEEE Sensors Letters,* vol. 4, no. 5, pp. 1-4.

[9]   M. Khan, and R. Astya, "Face detection and recognition using opencv", *2019 International Conference on Computing, Communication, and Intelligent Systems (ICCCIS),* 18-19 October, Greater Noida, India, 2019, pp. 116-119.

[10]   Xui. Hui, and Qi. Miao, "Multimodal biometrics based on convolutional neural networks by two-layer fusion", *12th International Congress on Image and Signal Processing, BioMedical Engineering and Informatics (CISP-BMEI),* 19-21 October, Suzhou, China, 2019, pp. 1-6.

[11]   A.S. Falohun, S.O Amoo, and O.S. Rasheed, "Design of the parliamentary electronic voting response response system", *IOSR J. Comp. Eng.,* pp. 52-59, 2016.

[12]   K. Aggarwal, "Problems with the use of the online voting system in India", *Int. J. Eng. Comput.,* vol. 6, no. 5, pp. 5285-5288, 2016. [http://dx.doi.org/10.4010/2016.1294]

[13]   D.R. Ajmera, and R. Gautam, "International journal of advanced research in computer science and software engineering", *Int. J. Adv. Res. Comput. Sci. Softw. Eng.,* vol. 4, no. 1, pp. 584-587, 2014. [http://dx.doi.org/10.26483/ijarcs.v8i5.4021]

[14]   V.S. Grandfather, A.L. Siridhara, R. Karthik, K.S. Rao, and G. Bhavana, "Electronic operating vote working user vote", *Int. J. Civ. Eng. Technol.,* vol. 8, no. 7, pp. 214-218, 2017.

[15]   P. S. Balakrishnan, "Aadhar based FQ voting machine", *Imp.J. Interdiscip.Res.,* vol. 3, no. 3, pp. 1247-1251, 2017.

[16]   G.R. Bhatia Vaibhav, "GSM-based electronic voting machine design that can track voters. BIJIT - BVICAM's", *Int. J. Inf. Technol.,* vol. 7, no. 1, 2014.

[17]   R.J. Boram, *"Shoup corporation, bryn mawr",* U.S Patent 4641240

[18]   D. Card, and E. Moretti, "Does voting technology affect election results? Touch screen touch and 2004 presidential election", *Rev. Econ. Stat.,* vol. 89, no. 4, pp. 660-673, 2007. [http://dx.doi.org/10.1162/rest.89.4.660]

[19]   S. Chakraborty, S. Karmakar, and S. Dey, "Design of secured wireless real time electronic voting machine", *Int. J. innov. res. elect. electro. instrum. cont. eng.,* vol. 3, no. 9, 2015.

[20]   P. Chugh, and P. Dimri, *Review of existing india voting system and mixed design using biometric protection in voting verification process.,* pp. 28-27, 2015.

CHAPTER 9

IoT-Based Automated Decision Making with Data Analytics in Agriculture to Maximize Production

A. Firos[1,*], Seema Khanum[1], M. Gunasekaran[2] and S.V. Rajiga[3]

[1] *Department of Computer Science and Engineering, Rajiv Gandhi University, Rono Hills, Doimukh-791112, India*

[2] *Department of Computer Science, Government Arts College, Salem-636007, India*

[3] *Department of Computer Science, Government Arts College, Dharmapuri-636705, India*

Abstract: This study presents a technique for solving the real-time decision-making difficulty in farming due to sudden changes in situations like atmospheric changes, monsoons, pest attacks, *etc.* The future of farming technologies is collecting and analyzing big information in agriculture to maximize effectiveness that is operational and minimize work costs. But there tend to be more styles to comprehend with all the IoT, therefore the Internet of Things will touch many more companies than simply farming. This study is focused on adapting the capability of IoT for data collection of features of crops and for automated decision-making with data analytics algorithms.

Keywords: Agriculture, Cost management, Decision making model, Data analytics, IoT, Pest classification, Pest detection.

INTRODUCTION

The farming industry will become more important than ever before in the next years. The UN projects that by 2050, there will be 9.7 billion people on earth, which may result in a 69% rise in agricultural manufacturing on a global scale. Ranchers and rural businesses are adopting the Internet of Things (IoT) for research and enhanced creative capabilities to meet the growing need [1 - 5].

The UN predicts that by 2050, there will be 9.7 billion people on earth, with agriculture activity increasing by 69% of total production between 2010 and 2050 [6]. Ranchers and rural businesses are embracing the Internet of Things for research and enhanced creative capabilities in order to meet this requirement. The UN predicts that by 2050, there will be 9.7 billion people on earth, with agriculture activity increasing by 69% of total production between 2010 and 2050

---

* **Corresponding author A. Firos:** Department of Computer Science and Engineering, Rajiv Gandhi University, Rono Hills, Doimukh - 791112, India; E-mail: firos.a@rgu.ac.in

S. Kannadhasan, R. Nagarajan, N. Shanmugasundaram, Jyotir Moy Chatterjee & P. Ashok (Eds.)

[6]. Ranchers and rural businesses are embracing the Internet of Things for research and enhanced creative capabilities in order to meet this requirement.

Actually, acquiring skills is nothing new. Before the Industrial Revolution, handheld tools had been the standard for more than 100 years [7 - 10]. Grain lifters, material composts, and the first farm tractor with internal combustion were all introduced in the 1800s. Fast-forward to the late 1900s, when ranchers began using satellites to organise their labour.

The IoT is expected to advance farming to the next level. Thanks to agricultural drones and sensors, smart farming is already becoming increasingly widespread among farmers. High-tech equipment is also swiftly becoming the norm.

This chapter's IoT applications in agriculture are described below, along with how "Internet of Things farming" might assist farmers in supplying the world's food needs in the years to come [11 - 15].

EXISTING TECHNOLOGICAL INTERVENTION IN AGRICULTURE

To improve the efficiency of their daily labour, planters have begun using high-tech advancements and technology in agriculture. For instance, sensors installed in fields enable ranchers to get precise maps of the topography and resources in a given area, despite factors like the causticity and heat of the soil. In order to predict environment designs for the days ahead, they may also access weather statistics.

Planters may use their smartphones to see their equipment, crops, and animals from a distance, as well as learn more about what their domesticated animals eat and produce. With the help of this invention, they can conduct accurate yield and animal projections.

Additionally, ranchers now use drones as a tool that is important for studying their holdings, doing field research, and producing ongoing data. As a significant example, John Deere, one of the most well-known brands in farming equipment, has started connecting its work vehicles to the Internet and has developed a method for handling information on the harvest yields of ranchers. Similar to self-driving cars, the organisation is leading the development of autonomous work vehicles that will free up ranchers to carry out other tasks and enhance their level of skill.

A few of these techniques go into precision farming, which is the most popular way to use satellite symbolism and other technology (especially sensors) to notice and record data with the aim of increasing production yield while minimising cost and protecting assets.

Brilliant nurseries use the IoT and related products to automatically offer a microclimate to trim assembly. These managed environmental conditions let ranchers have enough time for optimal adequacy while preventing conflicts with hunters and bad weather.

Crop splashing, water system, lighting, temperature, and dampness, which are only the tip of the iceberg, are just a few of the things that ranchers who utilise nursery that is reasonable monitoring frameworks may manage with the use of these bits of information from big data and research.

It is projected that the future of farming will include the use of IoT, agricultural sensors, and farming drones. Although cunning precision and cultivation are eradicating, they may just be the precursors to the supported use of innovation in the cultivation scene.

The rise of blockchain technology is moving towards the IoT and might have an impact on the farming sector because of its ability to provide organisations with important information about crops. Ranchers may use sensors to capture agricultural data that will be assembled into a blockchain and include attributes that are well-known, such as pH level, salt and sugar content, *etc.*

By 2023, agro sensors are expected to number around 12 million globally, according to insider knowledge. Additionally, innovation juggernaut IBM estimates that the typical ranch produces 50% of 1,000,000 informational leads each time, assisting ranchers in increasing advantages while improving returns.

Manure can be sprayed 40 to 100 times more quickly by robots than by humans. Given a significant share of the potential advantages of these IoT applications in the agribusiness, it makes sense that farmers are increasingly considering drones that are horticultural satellites for long-term farming. Ranchers may use drones to remotely monitor how far along their crops are in different stages of development. Ranchers may also use ingredients to revive plants that are sick or injured robots by spraying them with water. According to DroneFly, a robot can shower dung 40 to 50 times faster than doing it manually.

FARMING DATA COLLECTION WITH IOT

The Internet of Things (IoT) is increasingly prevalent in many aspects of our lives. We can see that the Internet of Things has a lot of potential for assisting us in our daily lives. IoT devices may gather different data from our everyday files, transfer them to a distant processing device, and have that device analyse it so that it can serve as the foundation for improving our lifestyles. It is not surprising that IoT devices have actually made their case in the horticulture world as well, enabling a more insightful approach to precisely screen animals and produce growth propensities, with new technologies constantly being developed and accepted by every contemporary industry. IoT is one of the trend-setting advancements that have the potential to alter agricultural guidelines, cost management, waste reduction, information organisation, product quality, and company performance.

IoT in the agro sector contains tools like soil sensors, cameras, and environment stations, along with other cutting-edge devices that collect data on cultivating activities. To turn the acquired data into useful experiences, computations are made using additional PC programmes. These insights may relate to a crucial nutrient in the ground, a serious threat from the annoyance, or the general well-being of domesticated animals.

Because of the continually growing population as a whole, there are also more mouths to feed on a regular basis. The most logical solution in this scenario is to use IoT in rural tactics. Horticulture requires substantial labour and is dependent on the environment and other necessary factors. Innovative development in the neighbourhood has been taking place for a time now. Robots will undoubtedly be used to watch ranch practises, and there is a trend towards the continued development of excellent farming practices.

Modern developments, such as IoT, have the potential to drastically alter the agricultural sector. The following are just a few of the five crucial ways that IoT is used to help farming:

It provides the capacity to calculate the production output, which helps arrange a cautious harvest transportation.

Cost Control and Reduced Spending

The possibility of harvest overproduction is reduced with proper management and yield supervision, resulting in less waste. The ability to anticipate gaps in agricultural development and animal welfare helps to reduce hosed output.

Information Control

Perhaps the agriculture sector has access to vast amounts of data, like weather patterns, soil quality, crop advancement examples, and dairy cattle health foundation that has been collected by smart agricultural sensors. This data keeps track of a company's development, employee performance, hardware efficacy, and other factors.

Expanded Product Quantity and Quality

By practising mechanisation, the general increase in employment in the farming sector jobs helps you take care of the whole assembly while also ensuring greater harvest quality.

Increased Business Effectiveness

This sufficiency is justified by the fact that IoT has led to the automation of a few cycles. One might automate many homestead duties, such as water system, preparation, or pest control, using clever IoT-related products.

According to the great majority of the benefits listed above, it is believed that the use of IoT in farming is finally leading to higher pay rates. The dairy industry uses IoT, which might sometimes be seen as a disruption. Wearable devices that track health-related data by tracking the individuals' daily activities have one of the applications in use. IoT is used by another device to analyse dense material. To make decisions about the creature's wellness, all of the provided data is collected and reviewed.

The farmers can take care of monitoring the health of their farmed animals thanks to IoT. It is beneficial to prevent the unfavourable death of pets. IoT-powered wearable devices are designed to monitor heart rate, blood pressure, wellness, and other parameters. Examples include wearable smartwatches and fit-piece bands. Any abnormality should be reported to the dog's owner. These data alert ranchers about animal-related problems.

Monitoring atmospheric conditions *via* IoT greatly aids farming. IoT delivers accuracy, fosters potential outcomes, and has an impact on projects being different. The deployed sensor equipment, both within and outside of the field, provides ongoing data about the environment. They analyse this data to forecast favourable environmental characteristics like humidity, precipitation, and temperature. IoT enables weather stations to automatically alter environmental conditions in accordance with the precise sets of bearings to develop clever nurseries. IoT sensors provide consistent data on the nursery's characteristics, such as temperature, brightness, moistness, and soil quality. By embracing IoT, the strategy as a whole becomes smarter and reduces individual interference while increasing accuracy.

One essential viewpoint for the network of harvest generation and inventory is grain capacity. In order to prevent waste or other problems, a sensor may detect the conditions in a grain additional room holder and, at that moment, convey information about the fill levels, temperature, and wetness in the storage space. Bugs are an obvious test for certified ranches. To counteract the unpleasant effects of this hazard, consistent action is required. In this aspect, IoT has great potential, and it also enables ranchers to more precisely manage crop welfare issues. These are IoT-enabled sensors that track the damage caused by bugs over time. Ranchers are using IoT-controlled bright irritation devices to prevent situations like the first-place scenario. Calculations using artificial intelligence powered by computers also aid in gathering sensor data and related environmental data. This information is supplemented by an increase in precise pesticide applications.

A general definition of accuracy farming is making informed judgements that are precise and effective. Brilliant applications, such as vehicle following, animal observation, stock checking, field observation, and so forth, improve the accuracy and oversight of the agribusiness approach. With careful cultivation, a rancher may assess the state of the soil in conjunction with several limits that affect functional production. It may help with establishing consistent operating conditions for the linked IoT to determine the supplement and water levels.

Farmland is increasingly being screened using drones. They aid in problem resolution and hence increase yield, which is often another growth area. Innovation undoubtedly alters our everyday lives in a variety of ways. Progress is made in the agricultural sector by improvements in horticultural hardware and the use of innovations like IoT for more advanced assembly. Due to this, it is now simpler than ever to complete projects that involve closely studying a firm. The ranch heads are presently using IoT to accurately monitor the standard of plants and the ongoing medical coverage of domesticated animals. Without a doubt, IoT-enabled wearable technology is a work that transforms the industry and speeds up executive procedures [16 - 21].

PROPOSED METHOD

Information, facts, and data have been collected for analysis or guidance. Since the rancher can control their production time which is guaranteed, data collection is the most essential component of accurate farming. There are several studies regarding precision farming in a crop that is quite different. In order to clarify and provide a solid example of how yield monitoring helps farmers recognise yield variability within an industry, make better variable-rate decisions, and develop a history of spatial industry information, a unique GIS database has been created. The practise of yield monitoring has evolved into one that is characteristic of grain mills and corn-soybean rotation systems from the past. This technology will be made available for purchase and tested on several crops, including sugarcane, potato, onion, sugar beetroot, tomato, hay, citrus and grapes. Accuracy farming has a number of parts that each play a crucial part in how systems work. The most often used aspect of precision agriculture is yield monitoring, which has been used to assess the yield variability of maize, soybeans, potatoes, tomatoes, onions, sugar beetroot, hay, oranges, grapes and sugarcane. With the use of yield monitoring systems, analyses of crop variety comparisons, yield damage reports, and industry efficacy are increasingly being done.

The initial step in the process of developing a crop might be soil preparation. The most extreme goal is to create a solid, weed-free seedbed for speedy crop germination and growth. Ploughing, which involves rotating and loosening the soil, is one of the main tasks necessary for major soil preparation. Additionally, one of the energies that are the most important for cultivation is getting the soil ready, which necessitates significant contributions. The risk of soil disintegration might increase due to the field's size. Today's precision cultivation equipment may help ranchers prepare their soil in a lot less time and with less fuel by enhancing the technique's accuracy, sufficiency, and supportability. Some data are generated by detecting data inputs. First of all, determining the hotness involves using a wind current approach since temperature control is concerned with radiation that is high when the temperature is rising and certain yields could be harmed. The temperature has an impact on how well-developed rural goods evolve in terms of improvement, germination, growth, and blooming. Then, at that time, water fume may be the major problem influencing how things proceed. The chance of the sickness is increased due to mugginess. Moisture may affect hydria push, which would close the stomata and allow the assimilation-dependent photosynthesis to shut down. The enhancement in yield is similarly impacted by more-finished soil water. Sort, age, stage, and climate are compared to determine the optimum available water and arrangements for things. A few important limits for water temperature include pH consideration, sogginess limits, electric conductivity, soil temperature, and others. Additionally, the light-dependent resistor [8] is essentially a conductive photo sensor. A remote sensor network (WSN), according to Gaikwad S.V. *et al.* [6], is unquestionably a framework made up of detecting, processing, and collaborating components that enable the brain to screen and comprehend in relation to the established bounds within the framework. Data collection, verification, reconnaissance, and clinical telemedicine are all common uses of WSN. Additionally, it is used in greenhouses and the water system framework for monitoring and managing parameters such as water flow, temperature, humidity, and other factors.

Fig. (**1**) shows the system model that this research used for crop data gathering and decision-making to advice farmers. You can use a variety of sensors to collect data about your neighbourhood, including heat sensors (LM35, DHT11, SHT75, SHT-3x DIS), moisture sensors (P-Hs220, DHT11, HH10D, CMOS chip, SHT-3 DIS), soil sensors (SN-M114, 10HS, DHT11), water-level sensors (pH measurement sensors), and light dependent resistors. The process begins with the detection of sensors, such as those that measure light, temperature, wetness, and soil moisture. The data will be properly gathered, processed, and saved in a database. The data will likely be prepared and a decision model (such as a Markov chain or decision tree) produced in the following step. Additionally, decision trees are an effective machine-learning method for category problems. The decision tree is built on a decision that has several levels or a tree-like structure. Each hub of the choice tree system offers a dual option that separates each course or a subset of courses from the other courses. Typically, the handling is finished by descending the tree until the leaf hub is reached. A hierarchical approach is really hinted at here. Additionally, this analysis helps the rancher identify their actual harvest time. Because of the change in hotness and stickiness, the system may warn and alert if anything happens in the workplace, prompting the rancher to take immediate action to address the problem, such as automating the water system. Fig. (**1**) illustrates how your decision model study's predictions of the crops that would provide the highest yield in a particular area over the long term might help ranchers increase their benefit return.

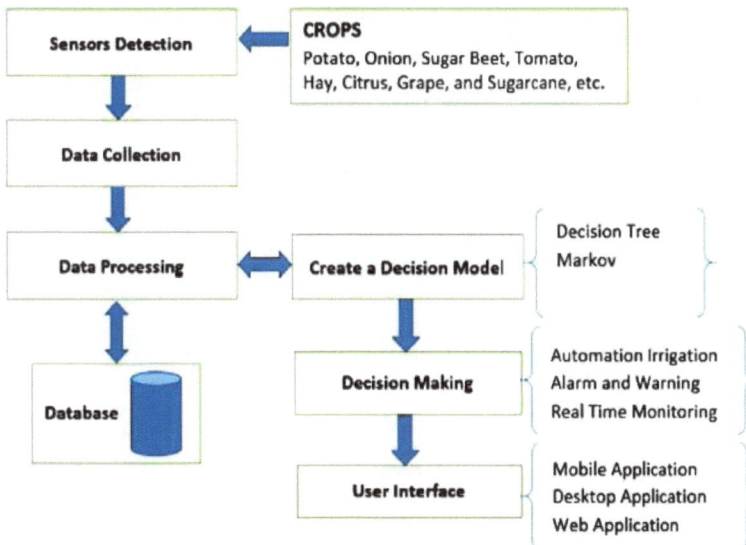

Fig. (1). System model of the study.

AUTOMATED DECISION MAKING WITH DATA ANALYTICS

The infrastructure, which allows devices to do the analytics, the supply of content, and a distribution method are the three kinds of provision needed to put the components of IoT data analytics together [12]. The enablers for end-user reach in crop feature data analytics are shown in the algorithm below. Each of these may be offered to stakeholders as a whole or customised to a configuration or necessity of an analytics engine.

The crop feature data history has to be pushed to analytics setup, for example, a Hadoop architecture using a flume script, as was previously indicated. For effectively gathering, aggregating, and transporting massive volumes of log data from several sources to a centralised data store, the Hadoop architecture, Apache Flume is a distributed, dependable, and accessible solution.

Finding Popular Pest Data, for Instance, 0x001 [African Bug]

for each tweet in the module

feature = extract(extract id, 0x001text)

end

for each feature

count_id = Count(id€0x001text)

endpopular_hashrtext = max(count_id)

Pest Classification

While T in C do

while words in tweet do

if word = = any phrase in dictionary_pest then

word_rating = d_r;

continue;

end

end

avg_rating = avg(word_rating)

if avg_rating ≥ = 0.0 then

Given pest is found positive

end

else if avg_rating< 0.0 then

Given pest is found negative

end

else

Given pestfoundis neutral(need further verification)

end

end

Pest Detection

Pest analysis is done using a dictionary-based method. A table is created to store the contents present in the dictionary of pests. In order to rate the tokenized words, the tokenized words have to be mapped with the loaded dictionary. We performed the left outer join operation on a table that contains id, word and dictionary table if the word matches with the pest word in the dictionary, then a rating is given to the matched word or else a NULL value is assigned. A hive table is created to store ID, word and then rating.

CONCLUSION

This research found that the IoT might be utilised to acquire localised data on precision farming. The farmer could quickly get the data that might be used to monitor his land in real-time. Provincial statistics include water level, temperature, stickiness, soil moisture, and light from various harvests (such as potatoes, onions, sugar beets, tomatoes, hay, citrus, grapes, and sugarcane). There are many different kinds of sensors that might be utilised to collect important data in the agricultural field. By providing predictions on the crops that may produce the most in a certain area, your choice of model research and choice of tree may help ranchers increase their edge of profit. Additionally, the calculations provide the result that foretells the preferred result that is advantageous.

REFERENCES

[1]   N. Ananthi, J. Divya, M. Divya, and V. Janani, "IoT based smart soil monitoring system foragricultural production", *IEEE Technological Innovations in ICT for Agriculture andRural Development (TIAR).* 07-08 April, Chennai, India, 2017, pp. 209-214.

[2]   Amandeep, A. Bhattacharjee, P. Das, D. Basu, S. Roy, S. Ghosh, S. Saha, S. Pain, and S. Dey, "IEEE annual information technology, electronics and mobile communication conference", In: *IEEE, Vancouver* BC, Canada, 2017, pp. 278-280.

[3]   M. Bodić, P. Vuković, V. Rajs, M. Vasiljević-Toskić, and J. Bajić, "Station for soil humidity, temperature and air humidity measurement with sms forwarding of measured data", *In 41st International Spring Seminar on Electronics Technology* 16-20 May 2018, Zlatibor, Serbia, 2018, pp. 1-5.

[4]   C. H. Chavan, and P. V. Karande, "Wireless monitoring of soil moisture, temperature & humidity using zigbee in agriculture", *Int. J. Eng. Technol.,* vol. 11, no. 10, 2014.

[5]   F. Ferrández-Pastor, J. García-Chamizo, M. Nieto-Hidalgo, and J. Mora-Martínez, "Precision agriculture design method using a distributed computing architecture on internet of things context", *Sensors,* vol. 18, no. 6, p. 1731, 2018.
[http://dx.doi.org/10.3390/s18061731] [PMID: 29843386]

[6]   S.V. Gaikwad, and S. G. Galande, "Measurement of NPK, temperature, moisture, humidityusing WSN", *Int. J. Eng. Res. Appl.,* vol. 5, no. 8, pp. 84-89, 2015.

[7]   I. Hedi, I. Špeh, and A. Šarabok, "IoT network protocols comparison for IoT constrained networks", *In Proceedings of the 2017 40th International Convention on Information and Communication Technology, Electronics and Microelectronics,* 22-26 May 2017,Opatija, Croatia, pp. 501—505.
[http://dx.doi.org/10.23919/MIPRO.2017.7973477]

[8]   R. Khan, I. Ali, M. Zakarya, M. Ahmad, M. Imran, and M. Shoaib, "Technology-assisted decision support system for efficient water utilization: A real-time testbed for irrigation using wireless sensor networks", *IEEE Access,* vol. 6, pp. 25686-25697, 2018.

[9]   A. Khattab, A. Abdelgawad, and K. Yelmarthi, "International conference on microelectronics", *28th International Conference on Microelectronics, ICM,* Giza, Egypt , pp. 201-204 year 2016.

[10]   B. Koch, and R. Khosla, "The role of precision agriculture in cropping systems", *J. Crop. Prod.,* vol. 9, no. 1, pp. 361-381, 2003.

[11]   R. Krishna Jha, S. Kumar, K. Joshi, and R. Pandey, "Field monitoring using IoT in agriculture", *2017 International Conference on Intelligent Computing, Instrumentation and Control Technologies (ICICICT),* 06-07 July, Kerala, India, 2017, pp. 1417-1420.

[12]   K. Nirmal Kumar, P. Ranjith, and R. Prabakaran, "Real time paddy crop field monitoringusing zigbee network", *International Conference on Emerging Trends in Electrical and Computer Technology (ICETECT),* 23-24 March 2011,Nagercoil, India, pp.1136-1140.
[http://dx.doi.org/10.1109/ICETECT.2011.5760290]

[13]   W.H. Nooriman, A.H. Abdullah, N. Abdul Rahim, and K. Kamarudin, "Development ofwireless sensor network for harumanis mango orchard's temperature, humidity, andsoil moisture monitoring", *In IEEE Symposium on Computer Applications & IndustrialElectronics* 28-29 April 2018, Penang, Malaysia, pp. 264-268.

[14]   G. Pavithra, "Intelligent monitoring device for agricultural greenhouse using IOT", *J. Agri.Sci. Food. Res.,* vol. 9, p. 220, 2018.

[15]   K. Prasad, A. Kumaresan, B. Nageshwaran, A. Kumaresan, and M. Kotteshwaran, "IoT based smart agricultural solutions to farmer enhanced with wifi technology", *Int. j. adv. res. basic eng.,* vol. 3, pp. 131-136, 2017.

[16]   P. Rajalakshmi, "IOT based crop-field monitoring and irrigation automation", *10th International*

Conference on Intelligent Systems and Control (ISCO), 07-08 January, Coimbatore, India, 2016, pp. 1-6.
[http://dx.doi.org/10.1109/ISCO.2016.7726900]

[17] T. Shakoor, K. Rahman, S. Nasrin Rayta, and A. Chakrabarty, "Agricultural production output prediction using supervised machine learning techniques", *In 2017 1st International Conference on Next Generation Computing Applications,* 19-21 July, Mauritius, 2017, pp. 182-187.
[http://dx.doi.org/10.1109/NEXTCOMP.2017.8016196]

[18] P. Srinivasulu, R. Venkat, M. Babu, and K. Rajesh, "Cloud service oriented architecture(csoa) for agriculture through the internet of things (iot) and big data", *In 2017 IEEEInternational Conference on Electrical, Instrumentation and Communication Engineering(ICEICE),* 27-28 April,Karur, India, 2017, pp. 1-6.

[19] Available from: http://internetofthingsagenda.techtarget.com/blog/IoT-Agenda/IoT-as-a-solution-forprecision-farming (Accessed on 4 October 2018).

[20] S. Veenadhari, D. Bharat Mishra, and D.C.D. Singh, "Soybean productivity modelling using decision tree algorithms", *Int. J. Comput. Appl.,* vol. 27, no. 7, pp. 11-15, 2011.
[http://dx.doi.org/10.5120/3314-4549]

[21] F. Viani, M. Bertolli, M. Salucci, and A. Polo, "Low-Cost wireless monitoring and decision support for water saving in agriculture", *IEEE Sens. J.,* vol. 17, no. 13, pp. 4299-4309, 2017.
[http://dx.doi.org/10.1109/JSEN.2017.2705043]

CHAPTER 10

An Indagation of Biometric Recognition Through Modality Fusion

P. Bhargavi Devi[1,*] and **K. Sharmila**[1]

[1] *Department of Computer Science, Vels Institute of Science, Technology and Advanced Studies (VISTAS), Chennai, India*

Abstract: One of the key predictions that had combined bio-sciences with innovation was bio-metrics, which represents a tool for security and criminology analysts to develop more accurate, robust, and certain frameworks. Biometrics, when combined with different combination techniques like feature-level, score-level, and choice-level combination procedures, remained one of the most researched technologies. Starting from uni-modular biometrics as unique marks, faces, and iris, they progress to multimodal bio-metrics. By presenting a similar investigation of frequently used and referred to uni- and multimodal biometrics, such as face, iris, finger vein, face and iris multimodal, face, unique mark, and finger vein multimodal, this paper will attempt to lay the groundwork for analysts interested in enhanced biometric frameworks. This comparative research includes the development of a comparison model based on DWT and IDWT. The method towards combining the modalities also entails applying a single-level, two-dimensional wavelet (DWT) that has been cemented using a Haar wavelet to accomplish the best pre-taking care of to eliminate disturbance. Each pixel in the picture is subjected to a different filtering operation in order to determine the peak signal to noise extent (PSNR). This PSNR analyses the mean square error (MSE) to quantify the disruption to hail before playing out the division of the largest dataset to the chosen MSE. In the most recent advancement, each pixel's concept is fixed up using the opposing two-dimensional Haar wavelet (IDWT), creating a longer image that is better able to recognise approbation, affirmation, and confirmation of parts. The MATLAB GUI is used to implement the diversions for this enhanced blend investigation, and the obtained outcomes are satisfactory.

Keywords: Authentication, Biometric, DWT, Fingerprint, Fusion, Iris.

INTRODUCTION

Fraud, information loss or exposure, and associated protected innovation are difficulties in today's PC-driven world. Security is essential to everything, including the concept of client verification. The verification measure provides

* **Corresponding author P. Bhargavi Devi:** Department of Computer Science, Vels Institute of Science, Technology and Advanced Studies (VISTAS), Chennai, India; E-mail: bhargavimowly@gmail.com

S. Kannadhasan, R. Nagarajan, N. Shanmugasundaram, Jyotir Moy Chatterjee & P. Ashok (Eds.)

security for monitoring, tracking, and accessing client individuality. There are many ways to authenticate a client, including information-based methods (such as a secret phrase or a unique proof number), token-based methods (like a security token, ATM card, or shrewdcard), and biometric methods. Despite the fact that passwords are straightforward, most implementations only provide the bare minimum of security. It is a pain dealing with several passwords for various frameworks. The use of security tokens or savvy cards is more expensive and calls for more specialised equipment and base support. The most secure form of validation has always been considered biometric. Biometric can only be stolen, lost, compromised, duplicated, or shared with much effort. They are also resistant to attacks on social structures. For their products, the organisations are participating in biometric verification. For businesses managing online business applications needing high levels of security, improving security for Web banking, ATMs, airports, legal requirement apps, and so forth, biometrics are essential. The foundation of ID, confirmation, and non-disavowal in data security is biometric verification [1].

Thus, a sensor module, an element extraction module, a coordinating module, and a selection module are included in the fundamental biometric structure shown in Fig. (**1**).

Fig. (1). A biometric system.

1. The sensor stage protects a person's biometric data, such as their fingerprints, by using a scanner to record a picture of their fingerprint.

2. The concentrations in which the supplied information is generated contain the values from the feature extraction step. For instance, the component extraction module of an iris framework would extract the mathematical forms (lines, spots, and so on) from an iris picture.

3. The component vectors are compared at stage 3.2's matching stage when the database's format creates the coordinating score. For instance, closeness estimates of the highlight vectors between the input face image and the information base face image will be logged and taken into account as a coordinating score.

4. Based on the coordinating score generated in the coordinating stage, the decision-making stages determine how the client's personality is regarded and is either recognised or rejected.

5. According to a few studies, the biometric validation framework based on the perception of the unimodal biometric format suffers from inadequate exactness caused by overly loud information, limited opportunities, non-distinctive and non-universal biometric characteristics, and execution restrictions [2].

6. In literature, the phrase "multi-biometric" is often used to denote biometric combinations. Therefore, in order to construct a multi-biometric framework, one must consider the three questions that have been covered in this article: (I) what to entangle, (II) when to circuit, and (III) how to combine. Choosing the many data sources to be combined, such as several computations or different modes, is part of what has to be intertwined. By analysing the various degrees of combination, or the many points in the biometric recognition pipeline at which data might be entangled, the question of when to combine is answered is shown in Table **1**. The phrase "step by step instructions to meld" refers to the combining approach used to combine the various data sources [3].

Table 1. Overview of major biometric traits.

S. No.	Traits	Advantages	Complications
1	DNA	Accuracy, discriminating power	Computational time, cost.
2	Finger Print	Accurate, unique, Consonant, smalldeposit Space.	Easily Hack able/Intrusive, dirt, dryness, finger Malposition, matchingis Calculation intensive.
3	Face	The process of Integrationiseasy, Touch-lesscapturing.	Lighting, camera angle, Twin'sproblem, storage Space and quality.
4	Iris	Stable ornament, high Degree of irregularity.	Hidden by eyelashes, lenses, Reflections, luminous illumination.
5	Palm	PrintDistinctiveness due to large area.	Expensive.
6	Ear	Less memory space, Consistent.	Obstruction by hair, ear Rings.
7	Signature	Templatescannotbe Stoleneasily, noninvasive.	High error rate, inconsistency.
8	Hand Geometry	Simple, easy to use, non-influence of Situation factors, less intrusive.	Not highly unique, not perfect for children, larger size of data.
9	Voice	Easy to use, less expensive	Alteration in voice due to sickness, can bemimicked, and can beoverheard if talkedaloud.

(Table 1) cont.....

S. No.	Traits	Advantages	Complications
10	Retina	Less memory space, noninfluence of illness	Intrusive, limited applications.

Note: International Journal of Innovative Technology and Exploring Engineering (IJITEE).
ISSN: 2278-3075, Volume-8, Issue-7S, May 2019, An Analysis on Biometric Traits.
Recognition Ebrahim A. M. Alrahawe, Vikas T. Humbe, G. N. Shinde.

A reliable biometric feature should satisfy each of the seven requirements listed below:

Universality: The trait should be present in every member of the target population;

Distinctiveness: the ability of the characteristic to sufficiently distinguish between any two persons;

Persistence: The property must be sufficiently invariant with respect to the coordinating rule throughout an arbitrary period of time;

Collectability: The characteristic should be easy to acquire or quantify;

Performance: Under a range of operational and environmental conditions, high recognition accuracy and speed should be feasible with constrained resources.

Acceptability: The biometric identification must be capable of open recognition, and the estimate tool must be secure.

Avoidance: It should be difficult to fake the feature *via* unethical means.

A multimodal biometric framework was created to address the drawbacks of a unimodal biometric framework. In a multimodal framework, proof and check measures are distinguished using at least two biometric modalities or other attributes. As a result, our architecture provides much stronger protection against parodying, as demonstrated in Table **2**. Any multimodal framework is meant to employ at least two different types of media. Any two modes together, or much more than just two, may be used. When compared to a mono-modular biometric framework, a framework using at least two modalities might be challenging to create and has perplexing functionality. The way it works is also unique; it makes use of a large number of sensors that may be combined to provide additional security (Multiple Instances) [4].

Table 2. Comparison of various biometric characteristics based on the table presented, High, medium, and low values are denoted by H, M, and L, respectively.

Biometric Identifier	Universality	Distinctiveness	Persistence	Collectability	Performance	Acceptability	Circumvention
DNA	H	H	H	L	H	L	L
Ear	M	M	H	M	M	H	M
Face	H	L	M	H	L	H	H
Fingerprint	M	H	H	M	H	M	M
Gait	M	L	L	H	L	H	M
Hand geometry	M	M	M	H	M	M	M
Iris	H	H	H	M	H	L	L
Palmprint	M	H	H	M	H	M	M
Signature	L	L	L	H	L	H	H
Voice	M	L	L	M	L	H	H

Note: IOSR Journal of Computer Engineering (IOSR-JCE) e-ISSN: 2278-0661, p- ISSN: 2278-8727Volume 15, Issue 1 (Sep. - Oct. 2013), PP 22-29 www.iosrjournals.org A ComparisonBasedStudy on Biometrics for Human Recognition Himanshu Srivastava Department of Computer Science & Engineering Roorkee Institute of Technology, Roorkee (U.K.), India.

PREVIOUS STUDIES

Since it relies on people's distinctive qualities for identification, biometrics—a mechanical science—has long been employed to verify human character in some way. The validity of biometrics depends on the fact that the characteristics used to identify evidence items are features that can't be lost or replicated; as a result, biometric methods are seen as being accurate, rapid, and trustworthy. The prior exams for fingerprint, iris, and multimodal combination processes are presented in this section.

Fingerprint and Iris Recognition System

• Multimodal fingerprint and iris biometrics with highlight level combo were introduced by Daniel *et al.* in 2012. The key features of the two modes were often connected and the framework was substantially implemented [5].

• Fingerprint and iris-based multimodal biometrics with element-level combinations based on repetition have been suggested by (Conti *et al.*, 2010). For experimentation, several information based test sets, such as 10 customers or 50 clients, are considered. In comparison to the FVC2002 DB2A finger imprint dataset and the BATH iris dataset separately, the Equal Error Rate is 2.36 percent for the FVC2002 DB2B fingerprint information base and 3.17 percent for the BATH iris information base, respectively [6].

• In 2019, Gunasekaran *et al.*, offered a DeepContourletDerivativeWeighted Rank (DCDWR) technique for biometric identification. The highlights were initially removed from photos of faces, fingerprints, and irises using a local derivative ternary pattern. The most significant attributes were kept in the histogram structure thanks to the distinct highlights, which enhanced the computationally complex nature of acknowledgment [7].

• PritiShivaji Sanjekar and colleagues in 2019, suggested Principal Component Analysis (PCA) to decrease the repetitive highlights. The combined methodology of highlight level combination is also suggested at the element level with coordinate score level and highlight level with choice level. On 100 distinct individuals drawn from three separate heterogeneous information sets, three approaches are examined. EERs of 1.2 percent, 1.15 percent, and 1.12 percent for the three combination conspiracies separately show good outcomes. For both (I) feature level linked with match score level and (ii) feature level united with decision level, 98.5 percent GAR is attained at FAR= 0.01 percent. The results of the trial showed that multimodal frameworks that incorporated the methods of highlighting at the coordinate score level and choice level outperformed highlighting at the element level alone [8].

• (Ahmed Shamil and colleagues, 2019). In this article, a multimodal combination techniques framework was constructed employing a combination of highlights that were subtracted from pictures of the iris and fingerprint using the GLCM computation. The present works recognised using iris and fingerprint were assessed prior to proposing the suggested acknowledgment framework. The results of the inquiry showed that the suggested framework, when combined with the KNN classifier, had a high exactness rate of up to 90%. The FAR, FRR, and completeness exactness rate were used to evaluate the framework [9].

• Walia *et al.* 2019 proposed a multimodal biometric system incorporating three complementary biometric features, namely the iris, finger vein, and fingerprint, based on an optimal score level fusion model [10].

• Swati K. choudhary *et al.* 2019 gives a summary of a number of biometrics used for verification. The scope of further testing in this field is discussed in this paper, with a particular emphasis on the difficulties associated with multimodal biometrics used in a variety of combination levels and computations. Notably, the importance of the multimodal biometrics format security problem is emphasized while applying a variety of ways to protect the vital human character resource [11].

• The identification of humans utilising two modalities, the iris and the fingerprint (K. Aizi *et al.*). Using a cutting-edge approach for the score range into zones of interest, we provide another combination method for the two modality selections: the dispersion of the ID scores generated by Hamming and Euclidean separations of the iris and fingerprint frameworks. Plotting the FAR, FRR, and EFAR error rates as indicated by different option limits ranging from 0 to 1 was done to test the two modalities [12].

• In 2019 (Sukhdevsingh *et al.*), This research used a multimodal biometric framework (FKP + Iris) after merging the results of many highlights. The highlights are extracted using the PCA. The coordination is finished by the NFNN classifier, which uses trial results to study the arrangement and coordinates it very precisely. Comparing the analytical results from PCA and fuzzy neural organisation, it can be shown that for a given problem, the fuzzy neural organisation provides greater accuracy than PCA and other classifiers [13].

• It presented a unique half-breed combination plot (Dwivedi, R *et al.* 2018) for a multi-biometric layout check that is guaranteed and is reliant on score and choice level blend is shown in Fig. (**2**). Dempster-Shafer's (DS) hypothesis of evidence and Mean-ClosureWeighting (MCW) weighting are used to combine scores from numerous matches pertaining to each approach in the combination choice level process [14].

CURRENT INDAGATION

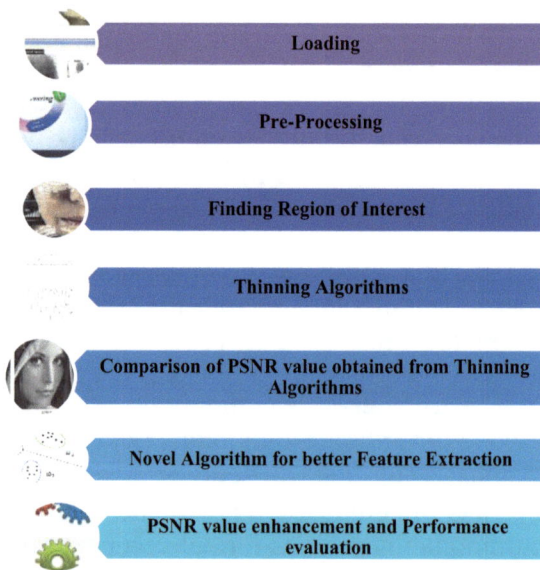

Fig. (2). Current Indagation.

CONCLUSION

This document describes the many technical procedures used with biometric technologies. For different pre-processing and post-processing approaches, biometric modalities have been explained using PCA analysis, classifier accuracy, and KNN algorithm. The fundamental approach taken into account for the investigation revolves around the use of the gray-scale co-occurrence matrix and the graph-theory technique. The present study focuses on pre-processing techniques to raise the image's PSNR value. The chosen photographs may benefit from substantial clarity and noise reduction as a result. The steps of post-processing, such as validation and identification, may be improved further by the processes of reducing noise and enhancing the picture.

REFERENCES

[1]   S. Ravi, and D.P. Mankame, "Multimodal biometric approach using fingerprint, face and enhanced iris features recognition", *2013 International Conference on Circuits, Power and Computing Technologies (ICCPCT)*, 20-21 March, Nagercoil, India, 2013, pp. 1143-1150.
[http://dx.doi.org/10.1109/ICCPCT.2013.6528884]

[2]   A.A. Altun, H.E. kocer, and N. Allahverdi, "Geneti algorithm based feature selection levelfusion using fingerprint and iris biometrics", *Int.J.Patt.Recog.Artif. Intell.,* vol. 22, no. 03, pp. 585-600, 2008.
[http://dx.doi.org/10.1142/S0218001408006351]

[3]   M. Singh, R. Singh, A. Ross, "A comprehensive overview of biometric fusion" *arXiv*, vol. 1902, p. 02919v1, 2019.

[4]   S. Narula Garg, R. Vig, S. Gupta "A Critical Study and Comparative Analysis of Multibiometric Systems using Iris and Fingerprints", *Int. J. Comput. Sci. Inf. Secur.*, vol. 15, no.1, 2017.

[5]   D.M. Daniel, and B. Monica, "A data fusion technique designed for multimodal biometric systems", *Proc. of 10th IEEE Int. Symp. on Electronics and Telecommunications* 15-16 November Timisoara, Romania, 2012, pp. 155-158.
[http://dx.doi.org/10.1109/ISETC.2012.6408095]

[6]   V. Conti, C. Militello, F. Sorbello, and S. Vitabile, "A frequency-based approach for features fusion in fingerprint and iris multimodal biometric identification systems", *IEEE Trans. Syst. Man Cybern. C,* vol. 40, no. 4, pp. 384-395, 2010.
[http://dx.doi.org/10.1109/TSMCC.2010.2045374]

[7]   K. Gunasekaran, J. Raja, and R. Pitchai, "Deep multimodal biometric recognition using contourlet derivative weighted rank fusion with human face, fingerprint and iris images", *Automatika,* vol. 60, no. 3, pp. 253-265, 2019.
[http://dx.doi.org/10.1080/00051144.2019.1565681]

[8]   S.S. Priti, B.P. Jayantrao, and R.C. Patil, "Institute of technology, shirpur, dist-dhule, india, multimodal biometrics using fingerprint, palmprint, and iris with a combinedfusion approach", *Int. J. Comp. Vis. Imag. Proces.,* vol. 9, no. 4, 2019.
[http://dx.doi.org/10.4018/IJCVIP.2019100101]

[9]   A. S. Mustafa, A. J. Abdulelah, A. K. Ahmed, "Multimodal biometric system iris and fingerprint recognition based on fusion technique" *Int. J. Adv. Sci.*, vol. 29, no. 03, pp. 7423-7432, 2020.

[10]   G.S. Walia, T. Singh, K. Singh, and N. Verma, "Robust multimodal biometric system based on optimal score level fusion model", *Expert Syst. Appl.,* vol. 116, pp. 364-376, 2019.
[http://dx.doi.org/10.1016/j.eswa.2018.08.036]

[11]   K.C Swati, and K.N. Ameya, "Multimodal biometric-based authentication with secured templatescannotbe", *Int. J. Image Graph.,* vol. 21, no. 2, p. 2150018, 2021.

[12]   K. Aizi, and M. Ouslim, "Score level fusion in multi-biometric identifification based on zones of interest", *J. King Saud. Univ. Comp. Inform. Sci.,* vol. 34, no. 1, pp. 1498-1590, 2022.

[13]   S. Singh, and C. Kant, "FKP and Iris based multimodal biometric system using PCA with NFNN", In: *Proceedings of International Conference on Sustainable Computing in Science, Technology and Management* Amity University Rajasthan: Jaipur India, 2019, pp. 26-28.
       [http://dx.doi.org/10.2139/ssrn.3358136]

[14]   R. Dwivedi, and S. Dey, "A novel hybrid score level and decision level fusion scheme for cancelable multi-biometric verification", *Appl. Intell.,* pp. 1016-1035, 2018.
       [http://dx.doi.org/10.1007/s10489-018-1311-2]

CHAPTER 11

A New Perspective to Evaluate Machine Learning Algorithms for Predicting Employee Performance

Dhivya R.S.[1,*] and **Sujatha P.**[2]

[1] *School of Computing Sciences, VISTAS, Thiruthangal Nadar College, Chennai, India*

[2] *Department of Information Technology, VISTAS, Chennai, India*

Abstract: Performance prediction is the forecast of future performance conditions based on past and present information. Forecasts can be made about companies, departments, systems, processes, and employees. This study focuses on assessing employee performance in terms of employee behavior, work, and growth potential. Organizations benefit when their employees perform well. Therefore, predicting employee performance plays an important role in a growing organization. To this end, we propose three machine learning algorithms: a support vector machine, a decision tree (j48), and a naive Bayes classifier. These can predict employee behavior in the workplace. Comparing the results, the Naive Bayes algorithm shows better results than the other two algorithms on the basis of metrics such as timeliness, error loss, and accuracy.

Keywords: Classification, Employee performance, Decision tree (j48), Naive bayes, Prediction, Support vector machine.

INTRODUCTION

Most businesses conduct quarterly or semi-annual performance reviews of their employees. It includes monitoring certain locations in need of development. And the yearly performance report shows the outcome. To increase attention and accomplish objectives, constant performance monitoring has been carried out recently on a weekly or monthly basis. For evaluating employee performance, there are several factors to consider. To anticipate employee performance, it takes into account important factors including the job satisfaction index, age, performance, number of employers, current years of employment, gender, job title, department, and total years of service. In this study, pilot tests are conducted

---

[*] **Corresponding author Dhivya R.S.:** School of Computing Sciences, VISTAS, Thiruthangal Nadar College, Chennai, India; E-mail: rsdhivya.vijay@gmail.com

S. Kannadhasan, R. Nagarajan, N. Shanmugasundaram, Jyotir Moy Chatterjee & P. Ashok (Eds.)

to choose an appropriate model for predictive comparison. According to experimental findings, Naive Bayes, Decision Trees, and SMO are the most accurate prediction models when compared to other models and will be employed in further studies.

LITERATURE SURVEY

The literature review that follows is highly useful in predicting customer behaviour and performance rating.

Muhammad TurkiAlshurideh *et al.* discussed raising customer turnover rates, dispersing categories, and client retention. He suggested, from a variety of angles, that appropriate categorisation be utilised to address problems with customer retention and customer-supplier relationships [1].

With the use of three-variable algorithms and complaints data, John Hadden *et al.* developed a novel method for forecasting customer attrition. They conducted a classification analysis on 202 records. Classification techniques include decision trees, regression, and neural networks, with neural networks achieving the greatest accuracy of 90% [2].

Ahmed Qureshi *et al.* used three methods based on their literature review since various authors used different algorithms depending on relevance and importance. P-values of certain characteristics were employed for data balancing, and the data set was resampled as well. The accuracy level was about 70%. The accuracy was enhanced to 75% using the decision tree when five more derived characteristics were used to improve it [3].

Manpreet *et al.* presented their results using graphs and other visual aids while using the J48 decision tree approach. The same Kaggle dataset that he had previously acquired was utilised. 'SGI' and 'data mining consultancy' both have copies of the dataset accessible on their websites [4].

By using KM in the HR system, LipsaSadath *et al.* described a data mining approach to forecast employee performance. The C4.5 approach exhibits greater accuracy in identifying various decision-making strategies on huge datasets in an automated and intelligent manner [5].

To estimate a student's final course grade, AI-Radaideh *et al.* employed a classification algorithm based on decision trees. They created a CRISP data-mining system for this purpose, which was utilised to mine academic student data [6].

A decision tree classifier was used by Surjeet *et al.* to conduct KDD in an educational setting and predict student performance. Additionally, it aided in the identification of dropouts, kids in need of particular attention, and students who need counselling and advice [7].

The classification methods Navie Naive Bayes, Decision Tree, and NBTree were compared by Alfisahrin *et al.* For this research, the 10 most relevant factors that contribute to liver illnesses were chosen. Higher accuracy is obtained using the NBTree method, while the quickest calculation time is obtained using Navie Naive Bayes [8].

A talent management challenge utilising C4.5 Data Mining methods was given by Hamidah Jantan *et al.* Based on prior information, they make predictions about the performance [9].

Tsai *et al.* used decision trees and neural networks as data mining techniques in mobile communications. They decided employing association rules to choose the most important variables to include during the preprocessing step. We can choose several metrics to assess model performance [10].

To forecast the performance of the new employee, Qasem *et al.* employed the ID3, J48 algorithms and Naive Bayes Classification approach. Data are gathered from many organisations to identify the most important performance-related criteria. According to the findings, an employee's job title has the greatest impact on their performance [11].

In order to categorise HR data, Yasodha *et al.* suggested a hybrid technique called CACC-SVM, which performs more accurately than conventional supervised algorithms. To identify a solution with greater accuracy, the different classification techniques such as Decision Trees, SVM, Neural Networks (NN), and closest neighbor algorithms were also contrasted [12].

On predicting employee turnover, Saradhi *et al.* compared several Machine Learning algorithms using employee attrition data and came to a conclusion that the work was highly beneficial for creating the best employee prediction model [13].

Decision trees were utilised by Kirimi JM *et al.* to determine the accuracy of employee performance using data from Kenya. The Institute for Public Management Development in Kenya paid close attention to the potential for creating new approaches for predicting employee performance and chose a superior one [14].

SVM, K-Nearest Neighbours (KNN), and DT were proposed in Human Resource Management by Desouki M. S. and colleagues. A multi-discipline academic research organization's Performance Appraisal (PA) outcomes were examined. Various data mining tasks have been used for prediction in order to enhance the assessment technique. DM tasks are more beneficial in human resource to improve the techniques of performance assessment [15].

In order to determine which Decision Tree approach had the highest accuracy for forecasting employee performance, V. Kalaivani *et al.* evaluated ID3, C4.5, Bagging, RandomForest, CHAID, CART, and Rotation Forest. Information is gathered from the organisation [16].

H. Jantan *et al.* classified employee accomplishment using the SVM approach. They employed several matrices in support vector machine approaches to get better outcomes in order to find talented employees based on their performance [17].

In order to analyse the mode of transportation, Ch.Ravi Sekhara *et al.* compared the Tree mode choice model, Multinomial Logit mode choice model, and Random Forest Decision. Since Delhi has significant levels of air pollution and considerable traffic congestion, 5,000 data points were gathered from residents of the city. The accuracy result for Random Forest was 98.96 percent, which was greater than previous results [18].

University students' suicidal ideation and associated risk variables were examined by Mohammadian-Khoshnoud *et al.* They then analysed the algorithms for logistic regression, random forest, and decision trees using criteria including receiver operating character, specificity, and sensitivity. Better outcomes are stated by decision trees [19].

In order to understand client retention, Bell *et al.* used ABMS (Agent Base Modelling and Simulation). To determine the causes of consumer behaviour and turnover from the current telecommunications business, the ABMS approach was investigated. Their choice of mobile device and location were also examined [20].

A decision tree or a random forest-based algorithm is held on the server, and the client receives input in the form of a feature vector classification. David J. Wu, *et al.* described two methods for confidentially assessing these algorithms. The initial protocol was based on semi-honest adversaries, where the client learned the model before the protocol was tested against malicious adversaries. As a result, both were able to process with multiple decision nodes in less time than the initial semi-honest protocols, which allowed them to realise that there had been an improvement in computation and bandwidth [21].

An analysis of data in Ghana's banking sector, which was affected by the financial crisis from 2015 to 2018, was carried out by Appiahene *et al.* Different categorization algorithms were used to anticipate the bank's performance and efficiency, which may also help to regain the trust of its customers. The C5.0 method produced superior results, with accuracy rates of 100%, 98.5% for random forests, and 86.6% for neural networks [22].

They compared decision tree and random forest on the four parameters accuracy, sensitivity, specificity, and the area under the ROC curve and came to a conclusion that the random forest has 71.3 percent sensitivity, 69.9 percent specificity, 71.1 percent accuracy, and 77.3 percent ROC than the Decision tree, which was used in the study by Habibollah Esmaily, *et al.* to identify Diabetes Mellitus using data mining from MASHAD (Mashhad Stroke and Heart Atherosclerotic Disorder) [23].

Ajay Kumar Mishra and colleagues performed a microarray analysis. It is an analysis carried out on microchips that have an array of small DNA elements that are attached for the purpose of gene expression. Two decision tree classifiers were produced from a dataset based on microarray data, trials were conducted using a methodical approach employing Weka tools, and the performance of the classifier was consistent throughout the dataset. We came to a conclusion that the attribute selection filtering procedures were at odds with one another and offered almost the same classification accuracy. They thus focused on statistical inference and discovered that the random tree-based attribute selection filter was better than the random forest filter [24].

In order to predict cardiac disease, Sujatha *et al.* examined Decision Tree, Naive Bayes, Random Forest, SVM, K-Nearest Neighbour, and Logistic Regression. When compared to other classification algorithms, Random Forest produces findings with a high accuracy of 83.52 percent [25].

METHOD

Tools Used

The WEKA workbench makes it easier to apply machine learning methods to a wide range of real-world issues. The datasets immediately use machine learning techniques. This tool performs classification, grouping, regression, and visualisation, and associates with the input data. It provides a setting for typical data mining issues. The UI for comparing different algorithms is user-friendly.

Use of Machine Learning Algorithms

The foundation of machine learning is the creation of computer programmes that can gather data and figure things out for themselves. This is incredibly efficient in all sectors because of the enormous quantity of data available. The resultant prediction model will be unparalleled, error-free, and time-saving when this is appropriately fed to the intelligent system and trained accordingly. As a result, the Kaggle dataset will be utilised to train the following ML model [4]. Today, classification and decision-making for complicated data are done using machine learning. Here, we forecast employee performance using three machine-learning techniques.

Bayes' Naive Classifier

The Naive Bayes classifier can handle a high number of variables in the supervised learning problem and only needs a few parameters that are linear in size. Statistical approaches for classification combined with supervised learning make it possible to train the naive Bayes classifier quite well. For classification problems, this probabilistic learning model is used. From the frequency count derived from the training data, the prior probability of attribute values and category-dependent categories are computed. By adopting an underlying probability model and calculating the likelihood of the outcome that aids in solving prediction difficulties, uncertainty may be recorded in a morally acceptable manner.

Support Vector Machine, or SMO

An approach called sequential minimum optimisation may solve quadratic programming issues that arise during support vector machine (SVM) training. SMO divides the issue into smaller issues and approaches each one methodically. As an internal loop, this avoids the laborious numerical QP optimisation. Since the amount of storage needed for a training set is linear, SMO can store a large training set. Because SVM assessment took up the majority of the SMO's processing time, SMO is quicker for linear SVMs. This approach transforms the nominal property to binary and substitutes any missing values globally. Additionally, by default, it normalises all characteristics.

Choice Tree

Weka is subjected to the decision tree (J48) method, and the decision tree serves as a representation of the recursive partition of instance space. A directed tree with no incoming edges is known as a root tree. Decision nodes with incoming edges are the remaining nodes. Each internal node separates the instance space into two or more subspaces depending on the discrete function of the input value.

In Fig. (**1**) the flow of work is explained through Schematic Diagram. Data collection, Analysis, Pre processing, algorithm used, and comparison are done.

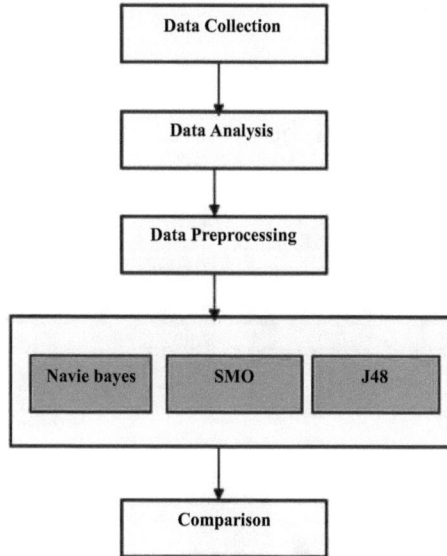

Fig. (1). Schematic Diagram.

Phases of Data

We break the lesson down into 5 sections. The first step in gathering performance statistics from the Kaggle website is data collection. After that, you may carry out exploratory data analysis as a second step. Data preparation, which includes attribute selection, is the third stage. The creation of the model is the fourth stage. Using three models, you may forecast employee performance using decision trees (j48), Naive Bayes, and SVMs. The last step is to evaluate the accuracy by contrasting these algorithms.

Data Collection

The performance data related to employees are collected from Kaggle website.

In Table **1** Employee Performance Data with their Corresponding Attribute, Instance, and class are detailed.

Table 1. Employee performance data.

Data Sets	Instance	Attribute	Class
Employee performance	1470	27	2

Data Analysis

Exploratory data analysis and results are shown below:

In Fig. (2), data are analysed and the most significant attributes are selected and compared.

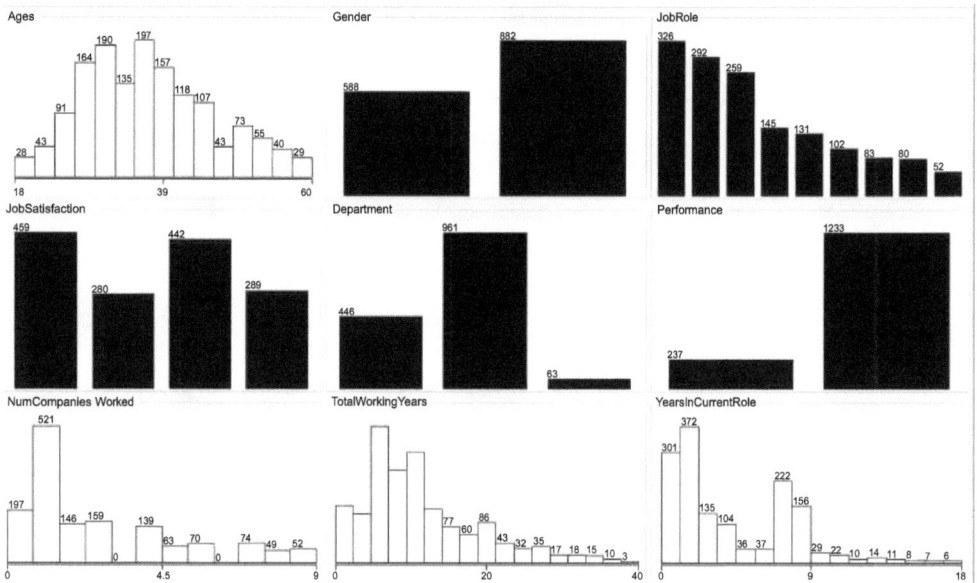

Fig. (2). Data Analysis.

Data Preprocessing

Preprocessing is essential to remove noisy data and select important data to process. Here, nine related attributes are selected from the employee background. Job satisfaction, age, performance, number of companies worked, years of current role, gender, level of work, department, and total years of work that are more important attributes.

In Table **2**, data are preprocessed and their details are classified according to their type, such as Unique, Missing, and Distinct.

Table 2. Employee details.

Employee Details				
Attributes	**Variable Type**	**Unique**	**Missing**	**Distinct**
Age	Numeric	0%	0%	43
Gender	Nominal	93%	0%	2
Job Satisfaction	Nominal	0%	0%	4
No of Companies Worked	Nominal	0%	0%	10
Years in current role	Nominal	0%	0%	19
Job role	Nominal	0%	0%	9
Performance	Nominal	0%	0%	2
Department	Nominal	0%	0%	3
Totalworkingyears	Numeric	0%	0%	40

Model Classification

Classification techniques are used to develop the Predictive Model. WEKA supports many classifiers, from which the following classifiers are selected. Algorithm J48, Naive Bayes and Support Vector Machines, attributes are ranked according to importance based on the information obtained.

Comparison of Models

After comparing the three algorithms using classification techniques, we can predict high accuracy of performance data.

RESULTS AND DISCUSSION

Static Evaluation

Employee data are statistically evaluated, and metrics such as min, max, and mean standard deviations are calculated. Each attribute is evaluated and shown below:

Tables **3** and **4** provide details about the statistical parameters and values regarding the given attributes.

Table 3. Employee details-1.

Employee Details				
Attributes	**Statistical Parameters**			
	Min	**Max**	**Mean**	**Std-Dev**
Age	21	40	30.5	10.367
Gender	2500	39999	6598.63	4818.36
Years in current role	0	18	4229	3.623
Total Working years	0	40	11.28	7.781
No of Companies Worked	0	9	2.693	2.498

Table 4. Employee details-2.

Employee Details				
Attributes	**Statistical Parameters**			
	Min	**Max**	**Mean**	**Std-Dev**
Performance	Label	Count	Weight	
	Yes	237	237	
	NO	1233	1233	
Department	Sales	446	446	
	Research& Development	961	961	
	Human Resources	63	63	
Job satisfaction	Four	459	459	
	Two	280	280	
	Three	442	442	
	one	289	289	
Job Role	Sales Executive	326	326	
	Research Scientist	292	292	
	Laboratory Technician	259	259	
	Manufacturing Director	145	145	
	Healthcare Representative	131	131	
	Manager	102	102	
	Sales Representative	83	83	
	Research Director	80	80	
	Human Resources	52	52	

Confusion Matrix

Table **5** shows details about the algorithms such as Naive Bayes, Support Vector Machine and J48 that are measured using the metrics F-Measure TP Rate, FP Rate and ROC rate. The Mean absolute Error, Kappa static, and time taken to run an algorithm are measured.

Table 5. Confusion matrix.

Classifiers	Confusion Matrix		
Naive Bayes Among 367 records, this algorithm predicted the employee performance as 31 showed "Good Performed" and 336 "Not performed". Predicted Yes (TP): 16 employees are truly classified as "Performed". Predicted No (TN): 52 employees are truly classified as "Not Performed". Actual Yes (FP): 15 employees are truly classified as "Performed". Actual No (FN): 284 employees are truly classified as "Not Performed".		yes	no
	yes	16	52
	no	15	284
Support Vector Machine Among 367 records, this algorithm predicted the employee performance as 18 showing "Good Performed" and 349 "Not performed". Predicted Yes (TP): 10 employees are truly classified as "Performed". Predicted No (TN): 58 employees are truly classified as "Not Performed". Actual Yes (FP): 8 employees are truly classified as "Performed". Actual No (FN): 291 employees are truly classified as "Not Performed".		yes	no
	yes	10	58
	no	8	291
J48 Among 367 records, this algorithm predicted the employee performance as 16 showed "Good Performed" and 351 "Not performed". Predicted Yes (TP): 11 employees are truly classified as "Performed". Predicted No (TN): 57 employees are truly classified as "Not Performed". Actual Yes (FP): 5 employees are truly classified as "Performed". Actual No (FN): 294 employees are truly classified as "Not Performed".		yes	no
	yes	11	57
	no	5	294

Tables **6** and **7** show matrics evaluation using Naive Bayes, SMO and J48 algorithm.

Table 6. Metrics evaluated using an algorithm.

Class	F-Measure	TP Rate	FP Rate	ROC Rate
Naïve Bayes	0.323	0.235	0.050	0.30
SMO	0.815	0.815	0.815	0.41
J48	0.765	0.820	0.734	0.50

Table 7. Mean absolute.

Class	Percent Correct	Kappa Statistic	Mean Absolute Error	Time
Naïve Bayes	81.74	0.23	0.26	0.00
SMO	59.42	0.00	0.30	0.03
J48	82.01	0.12	0.26	0.01

CONCLUSION

Employee data are included in this research, providing a comprehensive list of 1470 workers who have worked for the organisation and retired from it. Employees are assigned to one of many predetermined employment performance classes using employee keys and work-related documents. Two rulesets and three algorithms are generated using WEKA. The most accurate predictive model makes predictions about future instances of employee performance using the accuracy rate. Based on measures like accuracy, error loss, and timeliness, the performance assessment reveals that the Naive Bayes method outperforms the other two algorithms. The findings of this poll demonstrate that job performance and performance reviews are important indicators of employee performance. In

the future, it could be effective to use mask data to put the suggested model into practise.

REFERENCES

[1] M.T. Alshurideh, "Is customer retention beneficial for customers: A conceptual background?", *Journal of Research in Marketing*, vol. 5, no. 3, p. 382, 2016.
[http://dx.doi.org/10.17722/jorm.v5i3.126]

[2] J. Hadden, A. Tiwari, R. Roy, and D. Ruta, "Churn Prediction using Complaints Data", *World Acad. Sci. Eng. Technol.*, p. 19, 2006.

[3] A. Qureshi, "Telecommunication subscribers' churn prediction model using machine learning", *8th International Conference on Digital Information Management, ICDIM 2013,* 2013.

[http://dx.doi.org/10.1109/ICDIM.2013.6693977]

[4]   M. Kaur, "Churn Prediction in Telecom Industry Using R", *(IJETR)*, vol. 3, no. 5, 2015.

[5]   L. Sadath, "Data Mining: A tool for knowledge management in human resource", *International Journal of Innovative Technology and Exploring Engineering*, vol. 2, no. 6, 2013.

[6]   Q. A. AI-Radaideh, E.M. AI-Shawakfa, and M. I. AI-Najjar, "Mining student data using decision trees", *International Arab Conference on Information Technology (ACIT'2006)*, https://sites.google.com/site/ijcsis/ Yarmouk University, Jordan, 2006. International Journal of Computer Science and Information Security (IJCSIS), Vol. 17, No. 1, January 2019 39.

[7]   K.Y. Surjeet, B. Brijesh, and P. Saurabh, "Data mining applications: A comparative study for predicting student's performance", *International Journal of Innovative Technology and Creative Engineering*, vol. 1, no. 12, pp. 13-19, 2011.

[8]   S.N.N. Alfisahrin, and T. Mantoro, "Data mining techniques for optimatisation of liver disease clasification", *2nd International Conference on Advanced Computer Science Applications and Technologies (ACSAT 2013)*, Kuching, Malaysia, 2013.
[http://dx.doi.org/10.1109/ACSAT.2013.81]

[9]   H. Jantan, A. R. Hamdan, and Z. A. Othman, "Human talent prediction in HRM using c4.5 classification algorithm", *International Journal on Computer Science and Engineering*. 2 (08-2010), PP. 2526–2534 [D].

[10]   C.F. Tsai, and M.Y. Chen, "Variable selection by association rules for customer churn prediction of multimedia on demand", *Expert Syst. Appl.*, vol. 37, no. 3, pp. 2006-2015, 2010.
[http://dx.doi.org/10.1016/j.eswa.2009.06.076]

[11]   Q.A. Al-Radaideh, and E. Al-Nagi, "Using data mining techniques to build a classification model for predicting employees performance", *Int. J. Adv. Comput. Sci. Appl.*, vol. 3, no. 2, pp. 144-151, 2012.

[12]   S. Yasodha, and P.S. Prakash, "Data mining classification technique for talent management using SVM", the International Conference on Computing, Electronics and Electrical Technologies, 2012.
[http://dx.doi.org/10.1109/ICCEET.2012.6203768]

[13]   V.V. Saradhi, and G.K. Palshikar, "Employee churn prediction", *Expert Syst. Appl.*, vol. 38, no. 3, pp. 1999-2006, 2011.
[http://dx.doi.org/10.1016/j.eswa.2010.07.134]

[14]   JM Kirimi, and CA Motur, "Application of data mining classification in employee performance prediction", International Journal of Computer Applications, Volume 146 – No.7, July 2016.

[15]   M. S. Desouki, and J Al-Daher, "Using data mining tools to improve the performance appraisal procedure, HIAST Case", International Journal of Advanced Information in Arts, Science & Management Vol.2, No.1, February 2015.

[16]   V. Kalaivani, Mr. M. Elamparithi (2014), "An efficient classification algorithms for employee performance prediction", International Journal of Research in Advent Technology, Vol.2, No.9, September 2014 E-ISSN: 2321-9637.

[17]   H. Jantan, N. Mat Yusoff and M. Rozuan Noh (2014), "Towards applying support vector machine algorithm in employee achievement classification", Proceedings of the International Conference on Data Mining, Internet Computing, and Big Data, Kuala Lumpur, Malaysia, 2014 ISBN: 978---941968-02-4 ©2014 SDIWC.

[18]   C.R. Sekhar, M. Minal, and E. Madhu, "Multimodal choice modeling using random forest decision trees", *IJTTE Int. J. Traffic Transp. Eng.*, vol. 6, no. 3, pp. 356-367, 2016.
[http://dx.doi.org/10.7708/ijtte.2016.6(3).10]

[19]   M. Mohammadian-Khoshnoud, T. Omidi, J. Faradmal, and J. Poorolajal, *A comparison of random forest and decision tree for suicide ideation classification.*, 2020.
[http://dx.doi.org/10.21203/rs.3.rs-66839/v1]

[20]   D. Bell, and C. Mgbemena, "Data driven agent based exploration of customer behavior", *Simulation,* vol. 94, p. 003754971774310, 2017.
[http://dx.doi.org/0.1177/00375497717743106]

[21]   D.J. Wu, T. Feng, M. Naehrig, and K. Lauter, "Privately evaluating decision trees and random forests", *Proc. Priv. Enhan. Technol.,* vol. 2016, no. 4, pp. 335-355, 2016.
[http://dx.doi.org/10.1515/popets-2016-0043]

[22]   P. Appiahene, Y.M. Missah, and U. Najim, "Predicting bank operational efficiency using machine learning algorithm: Comparative study of decision tree, random forest, and neural networks", *Adv. Fuzzy Syst.,* vol. 2020, pp. 1-12, 2020.
[http://dx.doi.org/10.1155/2020/8581202]

[23]   H. Esmaily, M. Tayefi, H. Doosti, M. Ghayour-Mobarhan, H. Nezami, and A. Amirabadizadeh, "A comparison between decision tree and random forest in determining the risk factors associated with type 2 diabetes", *Spring,* vol. 18, no. 2, p. 412, 2018.

[24]   A.K. Mishra, and B.K. Ratha, "Study of random tree and random forest data mining algorithms for microarray data analysis", *Int. J. Adva. Elect. Comp. Eng.,* vol. 3, no. 4, pp. 5-7, 2016.

[25]   P. Sujatha, and K. Mahalakshmi, "Performance evaluation of supervised machine learning algorithms in prediction of heart disease", *2020 IEEE International Conference for Innovation in Technology (INOCON),* pp. 1-7, 2020.
[http://dx.doi.org/10.1109/INOCON50539.2020.9298354]

CHAPTER 12

Pre-process Methods for Cardio Vascular Diseases Diagnosis Using CT (Computed Tomography) Angiography Images

T. Santhi Punitha[1,*] and **S.K. Piramu Preethika[1]**

[1] *School of Computing Sciences, VISTAS, Pallavaram, Chennai, Tamil Nadu, India*

Abstract: The discipline of artificial intelligence (AI), which trains computers to comprehend and analyse pictures using computer vision, is flourishing, particularly in the medical industry. The well-known non-invasive diagnostic procedure known as CCTA (Coronary Computerized Tomography Angiography) is used to diagnose cardiovascular disease (CD). Pre-processing CT Angiography pictures is a crucial step in computer vision-based medical diagnosis. Implementing image enhancement preprocess to reduce noise or blur pixels and weak edges in a picture marks the beginning of the research stages. Using Python and PyCharm(IDE) editor, we can build Edge detection routines, smoothing/filtering functions, and edge sharpening functions as a first step in the pre-processing of CCTA pictures.

Keywords: Artificial intelligence (AI), Cardiovascular diseases (CVD), Coronary computed tomography angiography (CCTA), Coronary artery diseases (CAD), Stenosis.

INTRODUCTION

An essential prerequisite for originating that does extensive internal and external validation is data preparation. Computer vision is a field that deals with giving robots the ability to comprehend pictures from coronary computed tomography and angiography.

Segmentation is a difficulty in scientific image processing to express the severity of coronary artery plaque and is essential to study the contraction of heart arteries to prevent heart attacks. Typically, noise, poor resolution, and inadequate reproducibility have an influence on photographs. This study employs a variety of edge detection techniques, including canny, cvtColor, GaussianBlur, Dialate, and

* **Corresponding author T. Santhi Punitha:** School of Computing Sciences, VISTAS, Pallavaram, Chennai, Tamil Nadu, India; E-mail: santhipunitha.t@amjaincollege.edu.in

S. Kannadhasan, R. Nagarajan, N. Shanmugasundaram, Jyotir Moy Chatterjee & P. Ashok (Eds.)

Erode, as well as soothing and sharpening techniques, including filter2D, GaussianBlur, medianBlur, filter2D, and bilateral Filter.

BLOCK DIAGRAM FOR PREPROCESSING IN CCTA IMAGE

Edge Detection

The primary phase in the image processing is deciphering the contents of the picture. The method of edge detection on a picture is crucial for interpretation. Fig. (**1**) illustrates an edge detection approach for locating the edges of an item or area inside an image.

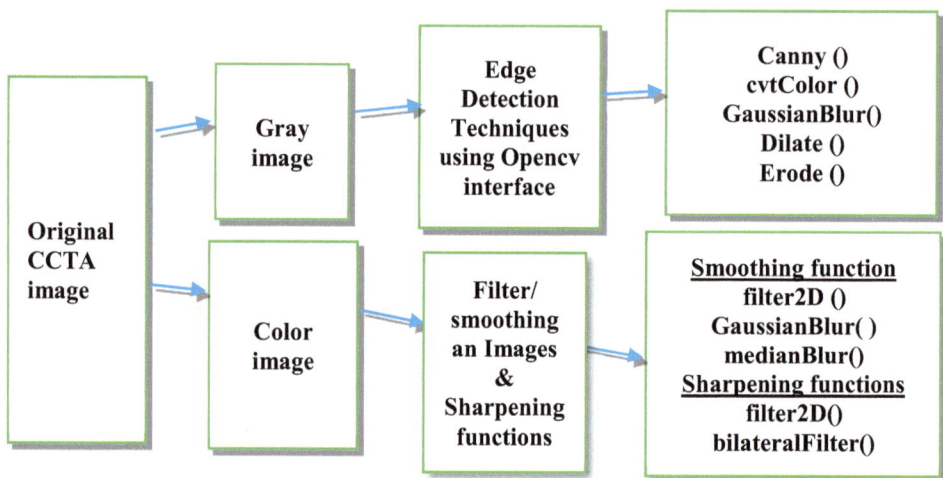

Fig. (1). Block diagram for preprocessing in CCTA image.

An edge is created by a fixed collection of connected pixels that span two endpoints. There are three different kinds of edges: a. horizontal; b. vertical; and c. diagonal.

A technique for dividing a CT picture into areas of discontinuity is called edge detection. It is used for picture morphology [1]. The viewer may see the image's structural elements in grayscale. It reserves the structural components of a CT picture while reducing the image data.

Edge Detection Operators:

There are two kinds of edge detection operators:

1. Gradient: It computes first-order derivations for digital cardiac pictures using algorithms like Sobel, Prewitt, and Robert.

2. Gaussian: In a CT picture, it computes second-order derivatives such as the Cannyedge detector and the Gaussian Laplacian.

To produce clear and precise edges, many edge-detecting algorithms have been created. The efficiency of image processing operations like picture segmentation and its retrieval will rise with accurate edge detection.

Steps to Perform Edge Detection in Python

1. Introduce Python Libraries

2. Select an image

3. Implement edge detection function

4. Execute the program

5. Save the image

Introduce Python Libraries

A process starting with a library needs to import three main libraries: Numpy, Matplotlib, and OpenCV.

Numpy, Matplotlib, and OpenCV are the three essential libraries that must be imported by a process beginning with a library.

Numpy is a Python package that supports expansive multidimensional arrays and a sizable number of high-level mathematical operations. We'll utilise the Matplotlib package to create static, animated, and interactive image visualisations in Python, and the OpenCV library for computer vision.

Select an Image

The imread ()function is used to read the image. The path of **the** image is passed as an argument. img = **cv2.imread ("d: /FFRCT_heartflow1_0.jpg")**

Implement Edge Detection Function

The canny edge detection model is another name for this OpenCV detection algorithm. Edge detection, visualisation, and result storage make up our function's three components [2]. Calling Canny is a way to conduct detection using OpenCV. The function's image is a parameter. That is, when we call the function, we provide the picture. The location array is necessary for the plot. After that, see both the original and edge-detection images [3 - 10]. The colour of the picture is altered using the cmap parameter. (Fig. **2**) illustrates how we converted them into grey in our situation.

Some of Edge Detection Functions are:

✓ **Kernel = np.ones((5, 5), np.uint8)**

✓ **imgGray = cv2.cvtColor(img, cv2.COLOR_BGR2GRAY)**

✓ **imgBlur = cv2.GaussianBlur(imgGray, (7, 7), 0)**

✓ **imgCanny = cv2.Canny(img, 150, 200)**

✓ **imgDialation = cv2.dilate(imgCanny, Kernel, iterations=1)**

✓ **imgEroded = cv2.erode(imgDialation, Kernel, iterations=1)**

Execute the Image

imshow() function is going to show us the plot that was created.

✓ **cv2.imshow(« clearimg », imgGray)**

✓ **cv2.imshow(« imgblur », imgBlur)**

✓ **cv2.imshow(« imgCanny », imgCanny)**

✓ **cv2.imshow(« imgDialation », imgDialation)**

✓ **cv2.imshow(« imgEroded », imgEroded)**

To Save the Result

The final part of the function will save the edge detected image and the comparison plot. *Imwrite* and *savefig* functions save images that are shown in Fig. (**3**).

✓ **cv2. Imwrite(« clearimg.jpg », imgGray)**

✓ **cv2. Imwrite(« img blur.jpg », imgBlur)**

✓ **cv2. Imwrite(« imgCanny.jpg », imgCanny)**

✓ **cv2.imwrite(« imgDialation.jpg », imgDialation)**

✓ **cv2.imwrite(« img Eroded.jpg », imgEroded)**

SMOOTHING/ FILTERING OF CCTA IMAGE

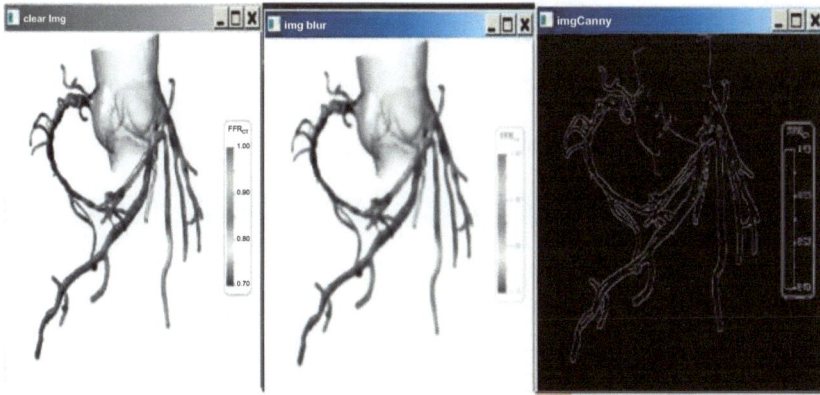

Fig. (2). Smoothing of CCTA Image.

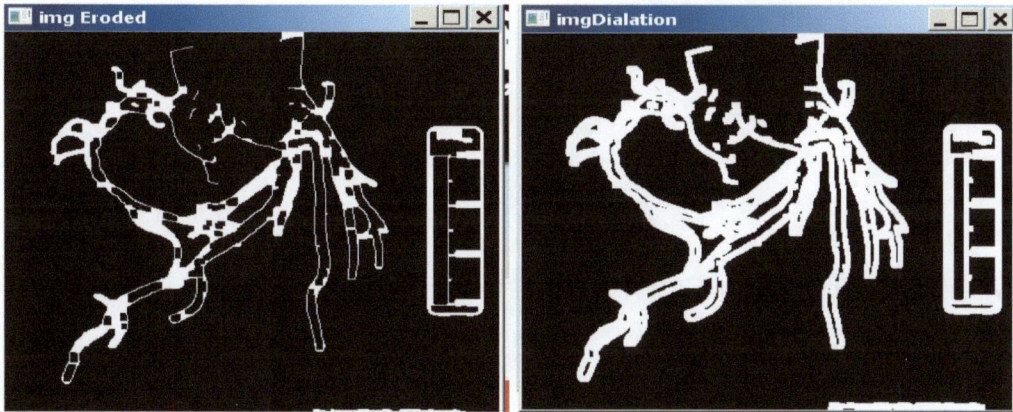

Fig. (3). Image Sharpening.

IMAGE SHARPENING

Filter Images using Convolution Kernels

The convolution kernel is a two-dimensional (2D) matrix for filtering images, the MXN matrix, where both M and N are odd integers (such as 3X3.5x5). The kernel performs math operations on each pixel of the image to achieve the desired effects such as image blurring/smoothing, image sharpening, and reduction of certain types of noise in the image.

BLURRING AN IMAGE

We can blur an image using filter 2D () function in OpenCV [8].

✓ kernel2 = np.ones((5, 5), np.float32) / 25

✓ img = cv2.filter2D(src=image, ddepth=-1, kernel=kernel2) is shown in Fig. **(4)**

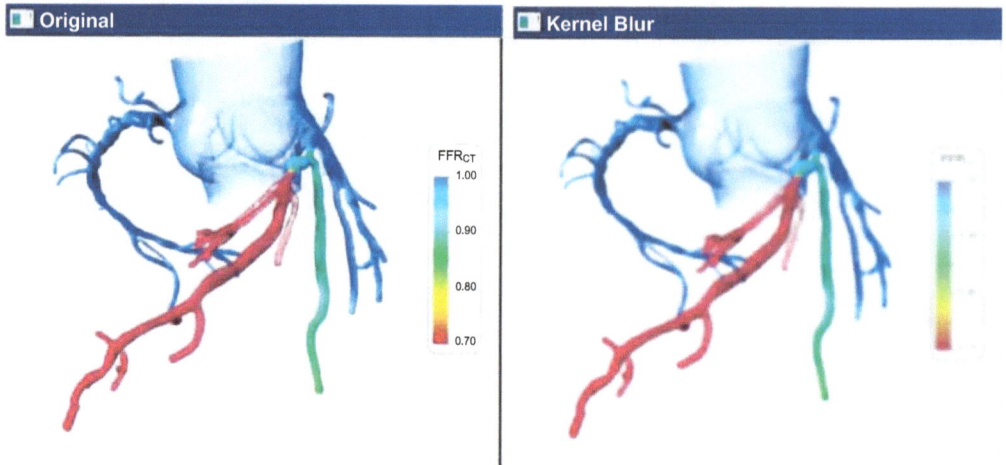

Fig. (4). Blur Kernel.

Gaussian Blurring to a CCTA Image

The ***gaussian filter*** performs a weighted average, as opposed to uniform average. This filter weights pixel values, based on their distance from the center of the kernel [9].

GaussianBlur(src, ksize, sigmaX[, dst[, sigmaY[, borderType]]])

gaussian_blur = cv2.GaussianBlur (src=image, ksize= (5, 5),sigmaX = 0, sigmaY = 0)

The final two arguments are sigma X (horizontal) and sigma Y (vertical), which are both set to 0 by default as shown in Fig. (**5**).

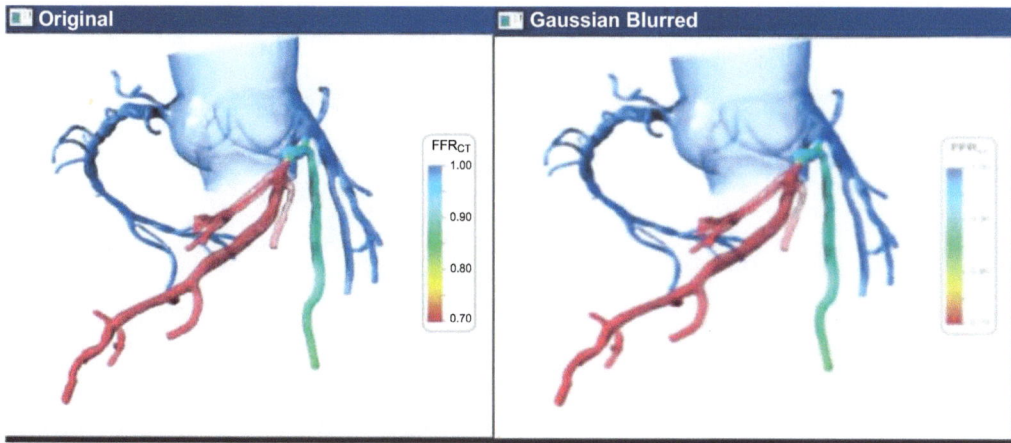

Fig. (5). Gaussian Bluered.

Applying Median Blurring to CTA Image

• In MedianBlurring, each pixel in the source image is replaced by the median value of the pixels in the kernel area as shown in Figs. (**6** and **7**).

median Blur (src, k size)

median = cv2.medianBlur(src=image, ksize=5) // herekernel size =5

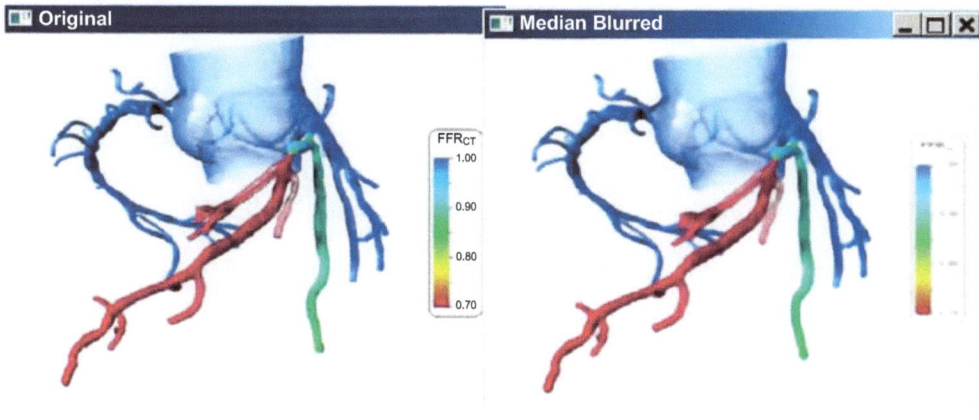

Fig. (6). Median Blurred.

sharp_img = cv2.filter2D(src=image, ddepth=-1, kernel=kernel3)

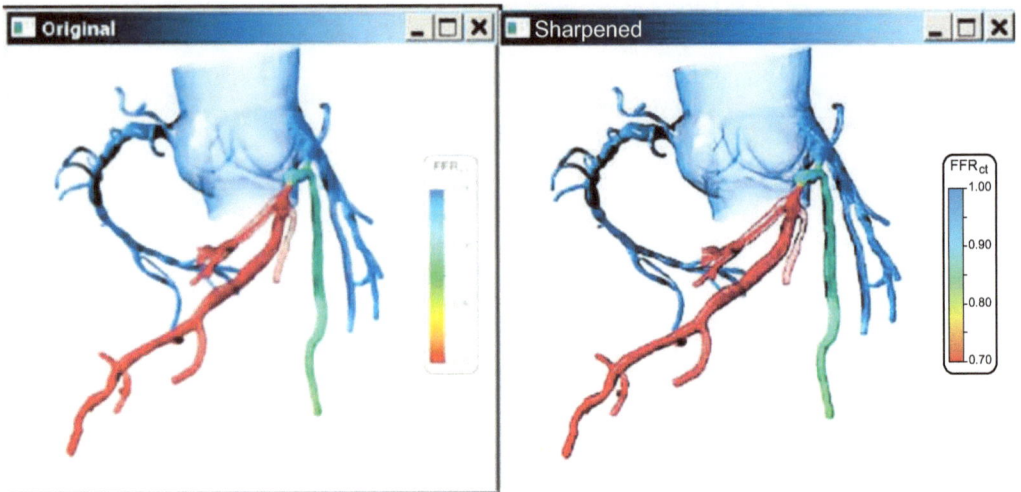

Fig. (7). Sharpened.

Sharpening an Image Using Custom 2D-Convolution Kernels

We can sharpen an image with a 2D convolutional kernel. First, define a custom 2D kernel, then use filter2D() function [10].

$$kernel3 = np.array([[0, -1, 0],$$
$$[-1, 5, -1],$$
$$[0, -1, 0]])$$

Bilateral Filter

While we blur the whole picture, some crucial information and sharp edges might be lost. However, this bilateral method applies the filters only to pixels in a neighbourhood that have comparable intensities, blurring them.

It may regulate the spatial size of the filter as well as how much the surrounding pixels are incorporated in the output of the filter. Based on their intensity and distance from the filtered pixel, this decision will be made.

bilateralFilter(src, d, sigmaColor, sigmaSpace)

Regions of more uniform intensity are blurred heavier, as they are not associated with strong edges.

The last(weighted) value for a pixel in the filtered image is a product of its spatial and intensity weights as shown in Fig. (**8**).

bilateral_filter = cv2.bilateralFilter(src=image, d=9, sigmaColor=75, sigmaSpace=75)

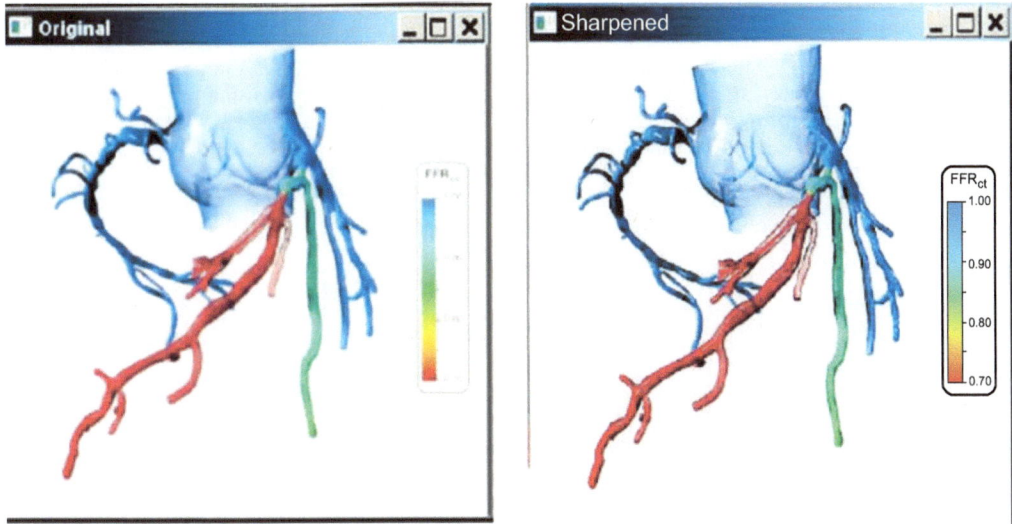

Fig. (8). Sharpened.

CONCLUSION

Python and the PyCharm IDE are used in this work to construct edge detection, smoothing, and filtering methods in order to sharpen the pixels and edges in CT Angiography pictures. By using segmentation techniques on the plaque-covered portion of coronary artery vessels, this study may be expanded in the future. We may also use quantum computing to enhance CT picture quality for precise prediction of cardiac plaque and estimate the proportion of blood vessel narrowing, preventing heart attacks at their earliest stages.

REFERENCES

[1] Available from: https://www.geeksforgeeks.org/image-edge-detection-operators-in-digital-image-processing/

[2] Available from: https://towardsdatascience.com/simple-edge-detection-model-using-python-91bf6cf00864Pop https://towardsdatascience.com/simple-edge-detection-model-using-python-91bf6cf00864

[3] J Liu, C Jin, and J Feng, "A vessel-focused 3d convolutional network for automatic segmentation and classification of coronary artery plaques in cardiac CTA", *International Workshop on Statistical Atlases and Computational Models of the Heart,* pp. 131-141, 2018.

[4] M. Pop, M. Sermesant, S. Zhao Jm Li, K. McLeod, and A.A. Young, Statisticalatlases and computationalmodels of the heart atrial segmentation and LV quantification challenges.*Proceedings of the 9ᵗʰ International workshop, STACOM 2018* Springer: Granada, 2018, pp. 131-41.

[5] M. Zreik, R.W. van Hamersvelt, J.M. Wolterink, T. Leiner, M.A. Viergever, and I. Išgum, "A

recurrent CNN for automaticdetection and classification of coronaryartery plaque and stenosis in coronary CT angiography", *IEEE Trans. Med. Imaging,* vol. 38, no. 7, pp. 1588-1598, 2019.
[http://dx.doi.org/10.1109/TMI.2018.2883807] [PMID: 30507498]

[6]　　Chen Chen, and Chen Qin, "Deep learning for cardiac image segmentation: A review", *Front. Cardiovasc.Med.,* vol. 7, p. 25, 2020.

[7]　　J.M. Wolterink, T. Leiner, B.D. de Vos, R.W. van Hamersvelt, M.A. Viergever, and I. Išgum, "Automatic coronary artery calcium scoring in cardiac CT angiography using paired convolutional neural networks", *Med. Image Anal.,* vol. 34, pp. 123-136, 2016.
[http://dx.doi.org/10.1016/j.media.2016.04.004] [PMID: 27138584]

[8]　　Available from: https://learnopencv.com/image-filtering-using-convolution-in-opencv/

[9]　　Available from: https://www.cnblogs.com/jasmineTang/p/14369043.html

[10]　　Available from: https://programs.team/convolution-image-filtering-in-opencv.html

CHAPTER 13

Implementation of Smart Wheelchair using Ultrasonic Sensors and Labview

N. Janaki[1,*], A. Wisemin Lins[1], Annamalai Solayappan[2] and E.N. Ganesh[1]

[1] *Vels Institute of Science, Technology and Advanced Studies, Pallavaram, Chennai, Tamil Nadu, India*

[2] *Sri Subramaniya Swamy Government Arts College, Tirutttani, Tamil Nadu, India*

Abstract: The patient-monitoring smart wheelchair system is a mobility aid for people with disabilities and continuously tracks the user's vital body metrics. Four interfaces—eyeball control, gesture control, joystick control, and voice control—have been created for wheelchair control in order to cater to various limitations. The image of the eyeball is captured using a camera. In order to make the necessary decisions based on the position of the eyeball, LabVIEW is used. The wheelchair movement can also be controlled by the other three modes. Anti-collision mechanisms are implemented using ultrasonic sensors. In the wheelchair, body temperature and heart rate monitoring provision is made. If any parameter is outside of a safe range, this system will notify the appropriate medical authorities and the wheelchair user's chosen individuals. The finished product is an innovative assistive technology that would simplify and lessen the stress in its user's life.

Keywords: MSP430, Patient monitoring, Python, Smart wheelchair.

INTRODUCTION

In India, there are persons with special needs. This demographic has a significant portion of physically challenged people. Independent movement is essential for people of all ages. Children who are unable to walk safely and independently are missing out on a crucial learning opportunity, which puts them at a developmental disadvantage in comparison to their peers who can walk safely and independently [1 - 5]. Although there are intelligent wheelchairs on the market, their price is rather expensive. The wheelchair in use is an attempt to make such cutting-edge assistive equipment affordable. Disabilities come in many forms, including paralysis, paraplegia, and muscular dystrophy.

* **Corresponding author N. Janaki:** Vels Institute of Science, Technology and Advanced Studies, Pallavaram, Chennai, Tamil Nadu, India; E-mail: janaki.se@velsuniv.ac.in

S. Kannadhasan, R. Nagarajan, N. Shanmugasundaram, Jyotir Moy Chatterjee & P. Ashok (Eds.)

There are four options available, including joystick mode, gesture control, ocular control, and voice control, depending on the kind of handicap. It has been made possible to monitor the heart rate and body temperature. The installation of ultrasonic sensors will offer obstacle detection and assure the wheelchair user's safety. When a person's bodily parameters rise over a certain level or the user hits the panic button, a GSM module has been installed that will send an alarm message to specific people so they may take the appropriate action.

As mobility assistance for those with disabilities, the patient-monitoring smart wheelchair device also continuously monitors the user's vital physiological data. Four wheelchair control interfaces—eyeball control, gesture control, joystick control, and voice control—have been created to cater to various limitations. The picture of the eyeball is taken with a camera. Depending on the location of the eyeball, LabVIEW is utilised to make appropriate decisions. The wheelchair may also be moved under control in three modes. For the implementation of the anti-collision system, ultrasonic sensors are used. The wheelchair has a feature for measuring body temperature and heart rate. If any parameter is outside of a safe range, the system will notify the appropriate medical authorities and the wheelchair user's chosen people. The finished product is a unique kind of assistive technology that would simplify and lessen the stress in its user's life.

The comfort of physically disabled people's lives is greatly aided by automation. In India, 41.32% of the population has a handicap of some type or another. The primary goal of this effort is to improve the lives of persons with physical disabilities and enable them to move about independently, free from the assistance of others or carers. It may be very helpful for those with muscular dystrophy, who are unable to move any portion of their bodies below the neck. They grow dependent on others due to muscular dystrophy, which causes them to feel bad and lose motivation. Such patients need the system to be self-sufficient in order to maintain their self-esteem. The existing systems for assisting the handicapped are expensive and out of the grasp of the average person. A middle-class family may also afford the system due to its high cost-effectiveness.

According to the goals of creating this mobility system, physically disabled persons would be able to move as they need to without depending on others. Patients with muscular dystrophy may be able to move independently by merely moving their faces. It may enable individuals who are completely paralysed live better lives.

The system is built using LabVIEW and the IMAQ vision package to run several digital image processing algorithms. This toolbox offers a full range of digital image processing and acquisition capabilities that boost system performance and require less human programming while producing better results faster [6-9].

SYSTEM OVERVIEW

The block diagram of the implemented system is shown in Fig. (1).

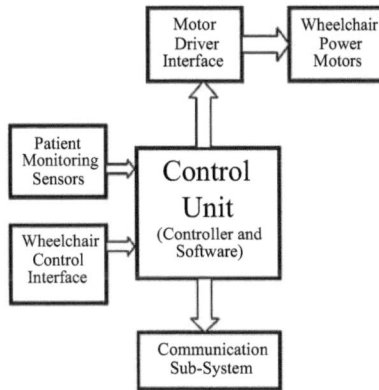

Fig. (1). Block diagram.

The system consists of the following sub-blocks.

Control Unit

MSP430 microcontroller, Python, and LabVIEW software make up the control unit. The control unit makes all of the choices and manipulates all of the data.

Wheelchair Power Motors

In order to move wheelchairs, DC motors are employed. High torque motors that can support the weight of a person sitting in a wheelchair are necessary because the real user will be seated in the wheelchair.

Motor Driver Interface

Those powerful motors cannot be driven by the controller. These motors need to be driven, hence a driver that can give the requisite current is needed.

Patient Monitoring Sensors

The wheelchair user's body temperature and heart rate are tracked using the temperature sensor LM35 and TCRT1000 module.

Wheelchair Control Interface

The wheelchair in use has four separate control schemes. These include joystick control, ocular control, audio control, and gesture control.

Communication Sub-System

This comprises the GSM module used to notify the proper authorities with alarm messages. This is accomplished by using a SIM 900 module. It can deliver AT instructions *via* a serial interface.

HARDWARE IMPLEMENTATION

MSP430 Microcontroller Series

There are multiple devices in the Texas Instruments MSP430 family of ultra-lo--power microcontrollers, each of which has a unique set of peripherals designed for a particular purpose. In portable measurement applications, the design is tailored to provide longer battery life *via* the use of five low-power modes. The system has a powerful 16-bit RISC CPU, 16-bit registers, and constant generators that help the code run as efficiently as possible. In less than one second, the digitally controlled oscillator (DCO) enables wake-up from low-power modes to active mode.

DC Motor

DC motors are used. The motors used are capable of producing adequate torque to propel the wheelchair under load when used with a spur gear. The DC motor utilised is shown in Fig. (**2**), and the motor specs are shown in Fig. (**3**), (Table **1**).

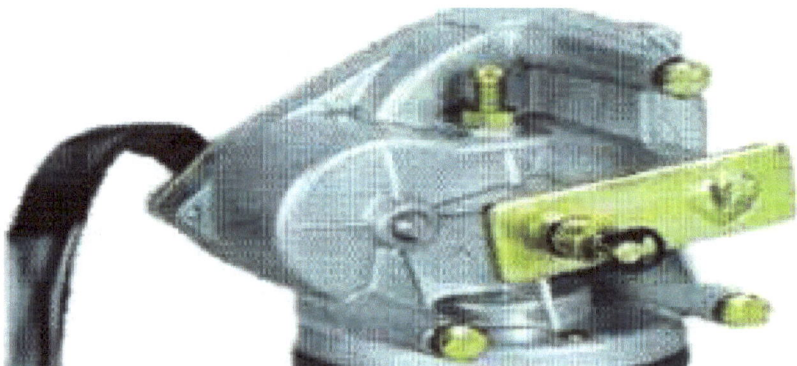

Fig. (2). DC Motor.

Table 1. Motor specifications.

Voltage Rating	17 NM
Power Rating	12 V
No Load RPM	60 rpm
On Load RPM	45 rpm
No Load Current	2.5 A
On Load Current	3.5 A

Fig. (3). Motor Driver Module.

Spur Gear

To attain the necessary torque value, spur gears have been employed to increase torque while decreasing speed. The spur gear assembly is shown in Fig. (4). The gear ratio was designed to enable the wheelchair to move under the necessary load.

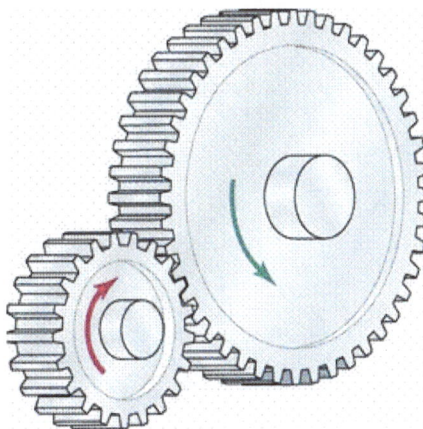

Fig. (4). Spur Gear.

Motor Driver Module

The drivér that manages the DC motors is this. Through the use of this module, the motors and ultimately the wheelchair are controlled by the microcontroller output. The driver used is shown in Fig. (5), showing the LCD used to display body temperature.

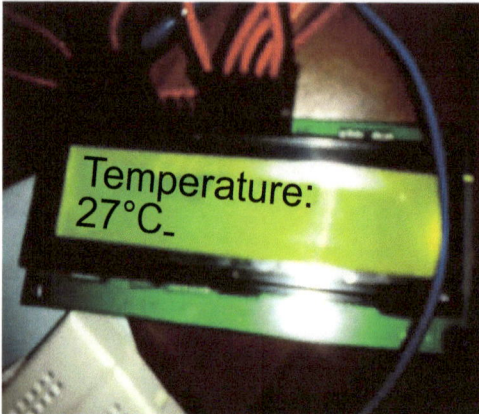

Fig. (5). Body Temperature Display.

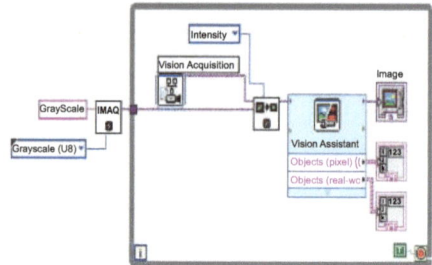

Implemented Wheelchair

Fig. (6) depicts the wheelchair as it is in use. This model has been equipped with gears in addition to the motors, batteries, and control box. The wheelchair may be operated by the user using the available controls. The wheelchair may be easily moved since it is collapsible.

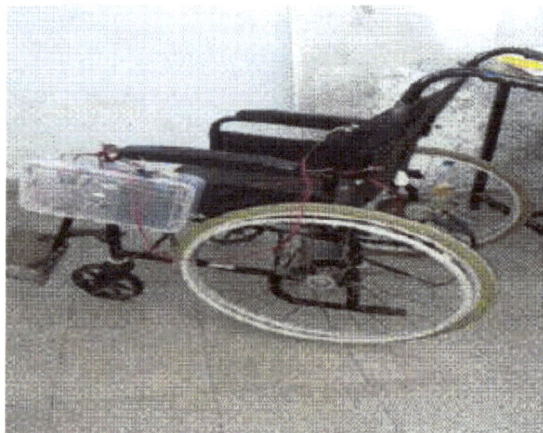

Fig. (6). Implemented Wheelchair.

HARDWARE IMPLEMENTATION

Eyeball Control

In this mode, the wheelchair is guided by the user's eye movements. Through the use of pattern recognition, this is accomplished. Pattern recognition is performed using Vision Assistant. The user's picture is matched with the template image. Real-time capture of the user's eye. The wheelchair is directed in the right direction by the microcontroller in accordance with the matching picture.

Audio Control

Wheelchair navigation is accomplished *via* user vocal instructions. Python is used to process the user's spoken command. Wheelchair control is done using a recognised command. The commands forward, reverse, left, right, and stop are employed. The graphical user interface (GUI) designed for the voice control mode is seen in Fig. (7). It will let the user know visually which command has been picked up.

Fig. (7). Audio Control GUI.

Joystick Control

This mode is provided for people who are paralyzed below the waist but are able to use their hands properly. Normal joystick operation is used to direct the wheelchair.

Gesture Control

This option is included for persons paralysed below the neck who can still use their neck adequately. The head motions are detected in this mode using a capacitive accelerometer, and the analogue data are converted into digital values using an ADC. The required action is taken by the microcontroller.

SOFTWARE IMPLEMENTATION

Python

An open-source programming language is Python. Voice control has been implemented using a speech module. The Tkinter module has been used to develop a GUI. Pyserial is used for serial communication. A certain character is sent serially to the microcontroller whenever a specific command is recognised. Depending on the received character, the microcontroller makes the appropriate choice.

LabVIEW

The system design platform and development environment for a visual programming language from National Instruments is called LabVIEW (Laboratory Virtual Instrument Engineering Workbench). Eyeball control mode is implemented using this programme. Eyeball tracking has been done using a Machine Vision module, and serial communication has been done using VISA.

CONCLUDING REMARKS

The final wheelchair that was built and put into use offers a contemporary, sophisticated assistive technology, which is a cost-effective answer. There are four ways to operate a wheelchair in the implemented system: eye-ball control, voice control, joystick control, and gesture control. For the implementation of obstacle avoidance and anti-collision mechanisms, ultrasonic sensors have been utilised. The wheelchair user's body temperature and heart rate are tracked, and if the readings rise beyond a certain typical level, the user-specified recipients will get an alert. A panic button is also available, and pressing it will sound an alert and send a message.

The patient-monitoring smart wheelchair system is a mobility assistance for persons with disabilities that continuously tracks the user's essential body metrics. Four interfaces—eyeball control, gesture control, joystick control, and voice control—have been created for wheelchair control in order to cater to various limitations. The picture of the eyeball is captured using a camera. In order to make the appropriate decisions based on the location of the eyeball, LabVIEW is employed. The wheelchair movement may also be controlled by the other three modes. Anti-collision mechanisms are implemented using ultrasonic sensors. There is a feature in the wheelchair for measuring body temperature and heart rate. If any parameter is outside of a safe range, this system will notify the appropriate medical authorities and the wheelchair user's chosen persons. The finished product is an innovative assistive device that would simplify and lessen the stress in its user's life.

Numerous methods exist for testing eye tracking. To determine if the idea is feasible, LabVIEW is used to simulate eye tracking for a smart wheelchair system. Compared to the most popular text-based programming languages, LabVIEW is a popular graphical programming environment that enables creating systems in an easy, block-based approach faster. In this study, LabVIEW will be utilised often for industrial automation, instrument control, and data collecting. As a consequence, this wheelchair makes it easier for the user to function effectively in areas with narrow entrances and ramps.

REFERENCES

[1]   R.C. Simpson, "Smart wheelchairs: A literature review", *J. Rehabil. Res. Dev.,* vol. 42, no. 4, pp. 423-436, 2005.
[http://dx.doi.org/10.1682/JRRD.2004.08.0101] [PMID: 16320139]

[2]   F Wallam, "Dynamic finger movement tracking and voice commands based smart wheelchair", *Int. J. Electr. Comput. Eng.,* vol. 3, no. 4, pp. 497-502, 2011.

[3]   T. Youcef, Smart powered wheelchair platform design and control for people with severe disabilities.*Computer Science & Artificial Intelligence Lab.* LIASD, University of Paris 8: Saint-Denis, 93526, France, 2005.

[4]   P. Dobhal, "Smart wheel chair for physically handicapped people using tilt sensor and IEEE 802.15.4 standard protocol", *Conference on Advances in Communication and Control Systems,* 2018.

[5]   R. Simpson, E. LoPresti, S. Hayashi, I. Nourbakhsh, and D. Miller, "The smart wheelchair component system", *J. Rehabil. Res. Dev.,* vol. 41, no. 3b, pp. 429-442, 2004.
[http://dx.doi.org/10.1682/JRRD.2003.03.0032] [PMID: 15543461]

[6]   Y. Hayashi, "Physical therapy for low back pain", *Japan. Med. Assoc. J.,* vol. 47, no. 5, pp. 234-239, 2004.

[7]   L. Fehr, W.E. Langbein, and S.B. Skaar, "Adequacy of power wheelchair control interfaces for persons with severe disabilities: A clinical survey", *J. Rehabil. Res. Dev.,* vol. 37, no. 3, pp. 353-360, 2000.
[PMID: 10917267]

[8]   J.D. Yoder, E.T. Baumgartner, and S.B. Skaar, "Initial results in the development of a guidance system for a powered wheelchair", *IEEE Trans. Rehabil. Eng.,* vol. 4, no. 3, pp. 143-151, 1996.
[http://dx.doi.org/10.1109/86.536769] [PMID: 8800217]

[9]   G. Pires, and U. Nunes, "A wheelchair steered through voice commands and assisted by a reactive fuzzy-logic controller", *J. Intell. Robot. Syst.,* vol. 34, no. 3, pp. 301-314, 2002.
[http://dx.doi.org/10.1023/A:1016363605613]

CHAPTER 14

Cryptography using the Internet of Things

T.R. Premila[1,*], N. Janaki[1], P. Govindasamy[1] and **E.N. Ganesh[1]**

[1] Vels Institute of Science, Technology and Advanced Studies, Pallavaram, Chennai, Tamil Nadu, India

Abstract: Cyberattacks on the power grid serve as a reminder that while the smart Internet of Things (IoT) can help us control our lightbulbs, it also runs the risk of putting us in the dark if attacked. Many works of literature have recently attempted to address the issues surrounding IoT security, but few of them tackle the serious dangers that the development of quantum computing poses to IoT. Lattice-based encryption, a likely contender for the next post-quantum cryptography standard, benefits from strong security guarantees and great efficiency, making it well-suited for IoT applications. In this article, we list the benefits of lattice-based cryptography and the most recent developments in IoT device implementations.

The Internet of Things (IOT) is a new technology that is anticipated to improve human lives. According to Cisco research, by 2020, there will be a vast array of IOT devices that will span every industry, including transportation, healthcare, and smart gadgets for every aspect of daily life. IOTs are improving user experience by making smart devices smarter and their services of high quality. The devices' unfettered access to the whole network makes the IOT's security issues more susceptible. The research paper will contribute to the presentation of a compiled report on the security issues with IOTs and the cryptographic techniques utilised to address them.

Keywords: Constrained devices, Digital signatures, Encryption, Lattice-based cryptography, Post-quantum cryptography.

INTRODUCTION

We now live in a global community thanks to the Internet, where emails from the US can be sent to China in less than a second, and real-time teleconferences link individuals from all over the globe. Even further, the Internet of Things (IoT) has an impact the physical environment in addition to how we transmit data (Fig. **1**). Smart home appliances, wearable technology that we use every day, driverless cars, and industrial control systems are just a few examples of how the Internet of

---

[*] Corresponding author T.R. Premila: Vels Institute of Science, Technology and Advanced Studies, Pallavaram, Chennai, Tamil Nadu, India; E-mail: premila.se@velsuniv.ac.in

S. Kannadhasan, R. Nagarajan, N. Shanmugasundaram, Jyotir Moy Chatterjee & P. Ashok (Eds.)

things has revolutionised how we live. It would be nearly impossible to purchase new technology in the near future that is not IoT-connected.

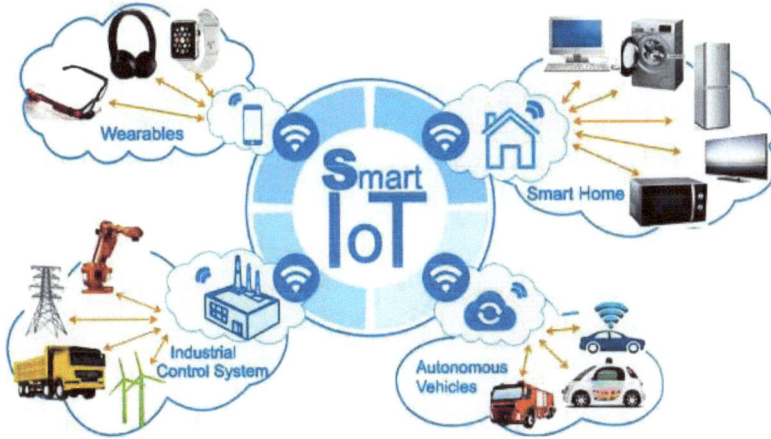

Fig. (1). Illustration of smart IoT applications.

By 2020, it is predicted that IoT technologies will have a several trillion dollar impact on the world's economy [1].

IoT privacy and security problems, however, loom over us constantly. Bruce Schneier [2], a security expert at Harvard University and the CTO of IBM Resilient, made the observation that IoT businesses are racing to make their devices cheaper and smarter but aren't paying much attention to security. The assaults on the Indian electrical system serve as a reminder that although smart IoT may help us regulate our lightbulbs if attacked, it might also leave us in the dark. Many works of literature have recently attempted to address the issues surrounding IoT security [3], but few of them examine the serious danger that improvements in quantum computing pose to IoT.

Despite the fact that there are still significant disagreements among scientists about quantum computers, many researchers are becoming more and more optimistic about the potential of large-scale quantum computers. The IBM Q system is a commercially accessible universal quantum computing system that IBM introduced in March 2017 as part of an industry-first project. It is intended for use in business and scientific applications. The commercially accessible 17-qubit processor is said to be at least twice as powerful as the 15-qubit universal quantum processor that is currently available to the general public.

Since they entail difficult-to-update platforms and systems, smart devices often utilised in smart IoT services are subject to the same, if not larger, quantum risks to cryptography. For instance, it's challenging to update embedded electronics in wearables and furniture, and the scaling issue with IoT devices makes the situation much worse. Therefore, while developing safe architectures and systems for the smart IoT, we need now take post-quantum security into account.

Post-quantum cryptography (PQC) has received attention recently thanks to Cheng *et al.* [4]. Lattice-based cryptography, a likely contender for the next PQC standard, benefits from strong security guarantees and great efficiency, making it well-suited for IoT applications. The benefits of lattice-based encryption and the most recent state-of-the-art IoT device implementations are the main topics of this article.

We first provide a quick overview of cryptography and the implications of quantum computers in the text that follows. Next, we discuss the benefits of lattice-based cryptography for smart IoT. After providing a high-level overview of lattice-based cryptography, we go into greater detail about the most recent lattice-based cryptography implementations on constrained devices. Finally, we discuss our thoughts on current issues and potential directions for further research into the use of lattice-based cryptography in IoT systems.

History suggests that innovation is driven by necessity, and as a result, contemporary science and technology have advanced in step with the desire to make people's lives simpler. As a result of the Internet of Things, the world is becoming increasingly interconnected every day. By 2020, roughly 31 billion linked gadgets will exist, according to Statista 2018. The increase in these gadgets' availability has made security the top priority.

For a very long time, both the makers and the customers have often ignored the security component of this vast network. It's imperative that we take a step back and consider security implications of this since our technology-dependent way of life is advancing us towards an Internet of Insecure Things.

The major topic of interest for researchers in this discipline has been security.

Multiple networked devices that are part of the Internet of Things (IoT) are constantly exchanging data and information with one another. We must understand the fundamental properties of security for IoT devices in order to secure that data:

• Confidentiality – We need to make sure that only authorised people have access to the information.

• Availability - Since many devices are linked, we need to ensure that each one receives the data it needs when it needs it.

• Integrity – We must guarantee the data's accuracy.

• Authentication – From the standpoint of the Internet of Things, this feature is crucial yet challenging to deploy. In the Internet of Things, there are several entities that are interconnected and provide a variety of functions.

• Heterogeneity – Each component of the network is unique in terms of its complexity, purpose, and even manufacturer. So, we also need to guarantee the network's heterogeneity.

The most significant phase is key encryption. The devices and other entities need a lightweight key management system to guarantee a secure connection.

Cryptography is a concept that has evolved through time as a result of data security and key transfer. By converting the data into an unrecognisable and unrelatable form, cryptography is a mechanism for protecting the data from unauthorised access. Dedicated cryptography algorithms must meet IoT criteria for being small in terms of size, memory footprint, power, and energy consumption.

In this study, we reviewed current research on the various phases of an IoT security solution. From lightweight cryptographic solutions to a comparison of various block cipher types, we thoroughly covered a flow of security measures. Additionally, we compared the most well-researched and reliable block cipher, Advanced Encryption Standard (AES), with several contemporary methods of security for Internet of Things devices.

CRYPTOGRAPHY AND QUANTUM COMPUTERS

Cryptography is used as a fundamental building block beneath all security protocols. Confidentiality, which calls for the inability of unauthorised parties to learn sensitive information, is the canonical implication of security. The most popular and straightforward method of achieving confidentiality is symmetric encryption. Alice and Bob, two communicating parties, share a secret key that can be used for both encryption and decryption. A third party cannot decipher the encrypted data from the ciphertext without knowing the secret key.

Symmetric encryption, which is a subset of symmetric-key cryptography, calls for two parties to share a common key. The difficulty of creating secret keys is one disadvantage of symmetric-key cryptography. This is typically accomplished through expensive secure channels like in-person meetings, the use of reliable couriers, or even quantum key distribution. These techniques are very expensive and challenging. In order to distribute cryptographic keys over insecure channels, asymmetric-key cryptography, also known as public-key cryptography, can be used. In public-key cryptography, Alice has two related keys; the private key is one, and the public key is the other. The private key, as its name suggests, is only known to Alice, whereas everyone else has access to her public key.

Anyone can encrypt a message using public-key encryption algorithms and send it to Alice using her public key. However, only Alice who is in possession of the secret key may decode. With the use of this feature, Bob may encrypt and send Alice a secret session key for a symmetric encryption method like AES. Alice may now create a secure channel with Bob using AES utilising the session key after successfully decrypting and receiving the AES key. Many security systems, including the Transport Layer Security (TLS) protocol, include what is known as hybrid encryption. Alice and Bob may also agree on a session key using the key exchange protocol *via* an unsecured channel.

Yet another issue appears. How can Bob, or anybody else, confirm that Alice indeed has the alleged public key but not Eve? This relates to the idea of cryptographic trust. There are often two options available. The Public Key Infrastructure (PKI) is one method, while Identity Based Encryption is another (IBE).

PKI is a technique that may link the identities of the owners of public keys to those keys. To demonstrate that a public key really belongs to an entity, a trusted certificate authority (CA) may issue a certificate to that entity. Informally, a certificate may be thought of as the CA's digital signature on the message "This public key belongs to Alice" using its private key. A digital signature of a message serves as the handwritten signature's digital equivalent, ensuring that the message was created by the signer (this relates to authentication in cryptography). The public key of the CA may be used by anybody to validate the signature of the CA and the certificate. Since CA is a reputable company, its public key must be well recognised. Since reputable CAs (such as governmental entities or international organisations) often have significant power and abundant resources to disseminate their public keys to the general public, this is a simple task to do.

The public and private key pair of an entity must be generated by a trusted authority in order to use the other IBE technique. However, no certificate is required. The public key of an entity in an IBE system may be anything, therefore an entity can use its identification as its public key, such as the name of a company or a person's email address, which can be readily confirmed by others. Users of the PKI system must confirm each certificate that CAs issue. As a result,

PKI requires complex public-key procedures, which are inherently unfriendly to IoT applications. IBE effectively lowers the cost of verifying the accuracy of public keys, which is advantageous in the context of IoT.

Modern cryptography rests its security on exacting proofs that guarantee security even under the most hostile conditions. The trust in the recognised difficulty of specific mathematical problems is reduced to the accepted security of almost all provably secure cryptographic primitives. Two well-known examples of this sort of issue are the discrete logarithm problem and the integer factorization problem (Elliptic Curve). They serve as the foundation for the popular modern cryptographic techniques RSA, Diffie-Hellman, and Elliptic Curve Cryptography (ECC). The most popular classical methods for factorization and discrete logarithm problems run on Turing computers with sub-exponential time complexity. However, both can be solved using Shor's quantum technique in a polynomial amount of time. A clear result is that our existing public-key cryptography system, including RSA and ECC, would be entirely broken if large-scale quantum computers became accessible. Therefore, it is crucial that we investigate alternative issues that are insurmountable for both conventional and quantum computers.

Grover's method, which offers a quadratic speedup for searching tasks over traditional algorithms, is another minor but widely impactful effect of quantum computing approaches. Grover's approach may be used to a variety of cryptanalysis techniques that call for using brute force. Grover's approach, for instance, helps speed up guessing the AES secret key. In general, one only needs to double the length of the key to obtain the same level of post-quantum security with respect to Grover's algorithm.

Government organisations, major enterprises, and university researchers all around the globe are fully aware of the quantum menace. The alternative technique, known as PQC, promises to provide cryptographic solutions that remain safe even when the attacker has access to powerful quantum computers. The National Security Agency (NSA) revealed its first intentions in 2015 for switching to quantum resistant encryption to safeguard sensitive data. The National Institute of Standards and Technology (NIST) published an open request for post quantum cryptography algorithms to be taken into consideration for standardisation in December 2016. When this article was written (in December 2017), the open call was over. The inaugural PQC standardisation conference is being organised by NIST and will take place in April 2018 so that the submitters may present and debate their proposals. In the Chrome web browser, Google is now testing post-quantum cryptography. The Tor project is attempting to create lattice-based key exchange protocols in order to achieve post-quantum security. Tor is a programme that shields its users from Internet monitoring.

WHY LATTICE-BASED CRYPTOGRAPHY?

To accomplish post-quantum security, a variety of ideas have been put forward, including lattice-based cryptography, multi-variate polynomial-based cryptography, hash-based signatures, and code-based encryption. In this essay, we concentrate on lattice-based cryptography. Lattice-based encryption, in our view, is well suited for smart IoT applications. First of all, lattice-based cryptography is well suited for Internet of Things (IoT) applications because of its excellent security assurances and great efficiency. Second, lattice-based cryptography has a broad range of applications that can support future developments in smart IoT services. Last but not least, of all subfields of post-quantum cryptography, lattice-based cryptography attracts the greatest attention. 28 of the 82 submissions for the most recent NIST call for post-quantum cryptographic algorithms are based on lattice, which is leading the field.

Lattice-based cryptography offers high levels of security. Although the underlying hard problems have been the subject of years of intensive study, neither a classical nor a quantum efficient algorithm exists for solving them. Lattice-based cryptography also benefits from a reduction in worst-case to average-case scenarios. Given the need for random keys, average-case intractability is a prerequisite for cryptography. Unless every instance of the underlying lattice problem is simple, the worst-case to average-case reduction essentially guarantees that lattice-based cryptography is secure on average. In terms of application, this worst-case reduction makes choosing parameters and creating keys for lattice-based cryptography much simpler. For example, the RSA cryptosystem is based on the hardness of integer factorization. However, this is the worst-case scenario. It is well known that the problem becomes essentially straightforward if the primes exhibit specific number-theoretic properties. Therefore, it's crucial to stay away from them when creating RSA keys. Unfortunately, we are unsure of how thoroughly these structures have been investigated. Lattice-based cryptography, in contrast, is based on hard problems that are typical in nature. Lattice-based cryptography only requires the uniform generation of keys after choosing the appropriate parameter size.

In contrast to the large integers used in RSA, lattice-based cryptographic algorithms operate over relatively smaller integers. The state-of-the-art computations for lattice-based algorithms primarily involve straightforward operations between matrices and vectors in a few rings or fields of low order. Lattice-based encryption really operates more quickly than RSA and can be used with low-power devices equipped with 8-bit microcontrollers. Lattice-based encryption implementations in recent times have already outpaced matching RSA systems by an order of magnitude. For instance, the state-of-the-art R-LWE based encryption implementation on an 8-bit AVR microcontroller can complete an encryption in 2 million cycles, whereas RSA-1024 implementations on comparable devices (which have a lower level of security and no post-quantum security) require more than 23 million cycles to complete the same task [5].

The performance of other post-quantum cryptography possibilities, such as code-based cryptography, may be even better in terms of computing efficiency, but these methods always call for bigger keys and ciphertexts. We emphasise that lattice-based cryptography is a good match for IoT applications because it strikes a balance among performance measures including key size, ciphertext and signature lengths, computational efficiency, and trust in security.

AN INTRODUCTION TO LATTICE BASED CRYPTOGRAPHY

Many efforts have been made in the research of PQC to seek out a replacement for widely used cryptosystems like RSA and ECC because the impact of quantum algorithms is mild for symmetric-key cryptography but devastating for current public-key cryptography. Lattice-based cryptography is one of them, and it shows a lot of promise. For a thorough overview of lattice-based cryptography, see Peikert [6].

The difficulty of solving certain geometric problems over high-dimensional lattices, like the shortest vector problem (SVP) and the closest vector problem, serves as the foundation for lattice-based cryptography (CVP). But if one has a solid foundation, these issues are simple to resolve. A strong basis consists of short, nearly orthogonal vectors, while a weak basis consists of long, generally pointing-in vectors. In Fig. (**2**), the blue points are lattice vectors—integer combinations of the basis B = [V1, V2]—which are coloured in blue. Based on almost orthogonal vectors, basis B′ is a good basis and basis B, which is coloured in green, is a bad basis. The SVP's goal is to identify the shortest nonzero vector, like vector V1′. The CVP is to identify the lattice vector T′ that is closest to an arbitrary vector in the space, such as the purple point T. For ease of visualisation, we use 2-dimensional (2D) examples, but SVP and CVP are simple in the 2D case. As the lattice's dimension increases, they become (exponentially) harder.

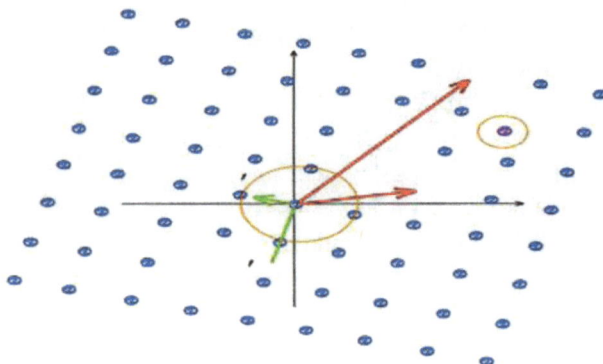

Fig. (2). Illustration of a 2-Dimensional lattice.

In Fig. (**3**), we present a list of the challenging issues in lattice-based cryptography. Worst-case hard problems are the SVP and CVP that were previously mentioned. Intractable situations are only guaranteed by problems of this nature. But generally speaking, the issue might be simple. Lattice-based cryptography has the significant benefit of having a large number of typical problems, such as the short integer solution (SIS) and learning with errors (LWE) problems. These are well-contained average problems that benefit from a worst-case to average-case reduction, which states that, unless the related problems on lattices are simple in all cases, SIS and LWE are difficult on average (for a random instance). The worst-case to average-case reduction makes it simple to create cryptographic schemes and demonstrate their security for lattice-based cryptography. In other words, one can build cryptographic primitives by working on conceptually straightforward average-case problems while also, in the worst case, gaining confidence in the constructions from low-level hard lattice problems.

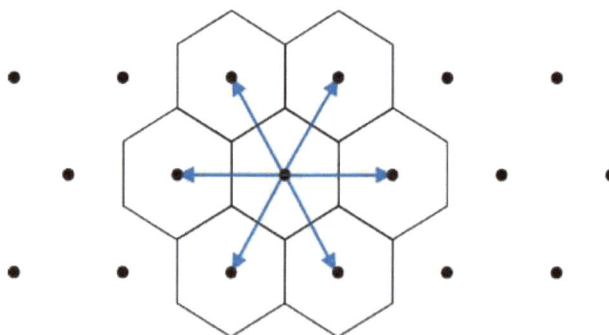

Fig. (3). Landscape of hard problems in lattice-based cryptography.

Solving linear equation systems is a simple way to describe the SIS and LWE problems. Finding short nontrivial (aside from all-zero solutions) integer solutions to a homogeneous linear system with a uniformly generated coefficient matrix is known as the SIS problem. The LWE problem asks you to locate the hidden vector S from a polynomially large sample set of AS + E, where AS is a uniformly generated matrix and E is a set of noises that you choose from a predetermined error distribution. Ring-SIS (R-SIS) and Ring-LWE are smaller rings of SIS and LWE, respectively (R-LWE). These ring variants can lower the memory and processing requirements for cryptographic schemes.

In reality, NTRUEncrypt is a lattice-based encryption method that has been standardised by the industry, in addition to primitives based on (R-)SIS and (R-)LWE. NTRUEncrypt's security is based on challenging puzzles over so-called NTRU lattices. Standard NTRUEncrypt does not have a worst-case reduction, but it has resisted assaults for 20 years (with an update in the parameters). Derandomizing LWE creates a different issue known as learning with rounding (LWR), where no error distribution is required to improve the performance of related techniques.

IMPLEMENTATIONS OF LATTICE-BASED CRYPTOGRAPHY FOR RESOURCE-CONSTRAINED DEVICES

The goal of seamlessly integrating everything makes it difficult to apply cryptographic algorithms on limited-resource devices like sensors and actuators. We examine the state of the art for lattice-based cryptography implementations on limited devices in this part. These implementations fall into two categories: software implementations on microcontrollers and hardware implementations on FPGAs.

Matrix-vector multiplication (schemes based on SIS and LWE) and polynomial multiplications are the primary operations of lattice-based cryptographic constructs (schemes based on R-SIS, R-LWE, and NTRU). The number theoretic transform (NTT) approach, a discrete Fourier transform (DFT) variation, may be used to improve polynomial multiplication utilised in R-LWE-based methods. Instead of the complex roots of unity used in DFT, the primitive root of unity modulo an integer N is employed in NTT. The complexity can be decreased from quadratic to quasi-linear by using the NTT to convert the conventional polynomial multiplication to point-wise multiplication. As suggested by Roy *et al.* for the R-LWE based encryption [7], the underlying algorithm can be changed to minimise the amount of NTT transformations required. It is also possible to use FFT optimization techniques.

The sampling from discrete Gaussian distribution is necessary for many lattice-based cryptographic constructions (LWE and R-LWE based) [8]. The discrete Gaussian sampler may be implemented using a variety of techniques, including rejection sampling, CDT sampling, Bernoulli sampling, and Knuth-Yao sampling. Due to the need for large pre-computed tables or a high precision computation of the exponential function, sampling from a discrete Gaussian is challenging.

The implementations of lattice-based proposals for key exchange protocols, public-key encryption, and digital signatures are covered in the sections that follow. (Table **1**), where the '/' inside a cell is used to separate figures for two different operations in the underlying algorithm, summarises the implementations on low-cost microcontrollers. For public-key encryption schemes, the two operations are encryption and decryption; for signature schemes, signing and verification; and for key exchange protocols, server-side computation and client-side computation. The memory used by the implementation is displayed in the "ROM." Table **2** provides a summary of the lattice-based cryptography hardware implementations on FPGAs.

Table 1. Software implementations of lattice-based cryptography on low-cost microcontrollers.

Schemes	Bit Security	Platform			Cycles	Time (ms)	ROM (KB)
		Device	CPU	MHz			
NTRUEncrypt [9]	128 (pre)	Cortex-M0 (XMC1100)	32-bit	32	588,044/950,371	18.4/29.7	9
R-LWEenc [5]	106 (pre) 46 (post)	ATxmega128	8-bit	32	796,872/215,031	24.9/6.7	6.5
R-BIN-LWEenc [10]	94 (pre)	ATxmega128	8-bit	32	1,573,000/740,000	49.2/23.1	2.7
		Cortex-M0	32-bit	32	999,000/437,000	31.2/13.7	5.6
IBE [11]	80 (pre)	Cortex-M0	32-bit	32	3,297,380/1,155,000	103.0/36.1	17
		Cortex-M4	32-bit	168	972,744/318,539	5.8/1.9	18.7
BLISS [5]	128 (pre)	ATxmega128	8-bit	32	10,156,247/2,760,244	317.4/86.3	18.4
NewHope [15]	128 (post)	Cortex-M0 (STM32F051R8T6)	32-bit	48	1,467,101/1,738,922	30.6/54.3	30.2
		Cortex-M4 (STM32F407VGT6)	32-bit	168	860,388/984,761	5.1/5.9	22.8

The IEEE P1363.1 standard has approved the NTRUEncrypt encryption as the first lattice-based encryption method. The creators of the NTRU cryptosystem have filed a patent for a variation that uses "product-from keys" to build the system quickly. However, they just recently (March 2017) declared that they will be making all of its NTRUEncrypt patents available to the public. Guillen *et al.* [9] investigated the viability of using NTRUEncrypt in limited devices (a Cortex-M0 based microcontroller).

Table 2. Software implementations of lattice-based cryptography on low-cost microcontrollers.

Schemes	Bit Security	Devices	MHz	Cycles	Time (μs)
R-LWEenc [7]	128 (pre)	V6LX75T	313	6,300 / 2,800	20.1 / 9.1
R-LWEenc [12]	128 (pre)	S6LX9	144	136,986	946
			189	66,338	351
BLISS [14]	128 (pre)	S6LX25	129	16,210	126.6
			142	9,835	69.3
IBE [11]	80 (pre)	S6LX25	174	13,958 / 9,530	80.2 / 54.8

By implementing an IBE scheme based on R-LWE, G uneysu and Oder [11] showed that IBE has become feasible even for embedded devices like Cortex-M microcontrollers and FPGAs. There are numerous R-LWEenc FPGA implementations; we present two of them. An FPGA implementation of R-LWEenc that was V6LX75T throughput-optimized was presented by Roy *et al.* [7]. With careful consideration of the parameters, P Oppelmann and G uneysu [12] presented an area-optimized implementation of R-LWEenc on S6LX9.

In terms of communication and computational costs, NTRUEncrypt is comparable to the R-LWE-based encryption scheme. The proven security of R-LWE-based encryption is one of its benefits, however a high-precision Gaussian sampler is required. On an 8-bit ATxmega128 microcontroller (32 MHz, 128 KB flash memory, 8 KB RAM), Liu *et al.* [5] provided a constant-time implementation of the R-LWE based encryption technique with a 46-bit post-quantum security level, with encryption and decryption times of 24.9 ms and 6.7 ms, respectively. By substituting a binary distribution for the Gaussian noise in R-LWE, Buchmann *et al.* [10] presented an encryption system that was implemented (R-BIN-LWEenc) on low-cost microcontrollers.

We recommend Howe *et al.* [13] for a thorough explanation of lattice-based signatures to the readers. The original proposals for lattice-based signature schemes that take advantage of CVP's hardness, like the GGH signature and NTRUSign, have been defeated. The most advanced signature method currently in use is called BLISS, which is based on R-LWE and has been shown to be secure in the random Oracle model. Because the deviation used in signature schemes is much larger than that in encryption schemes, discrete Gaussian sampling accounts for a larger budget in signature schemes than in encryption schemes. Liu *et al.* [5] and Poppelmann *et al.* [14] have implemented BLISS using state-of-the-art software and hardware, respectively. Ducas proposed a modification known as BLISS-B that can speed up key generation by a factor of 5 to 10 by lowering the repetition rate.

Google's post-quantum security experiments in Chrome have used NewHope, a post-quantum key exchange protocol based on R-LWE. A software implementation of NewHope for the 32-bit Cortex-M family of microcontrollers was presented by Alkim *et al.* [15]. Their implementation proved that lattice-based key exchange protocols are indeed promising candidates for post-quantum IoT security with a variety of generic and platform-specific optimizations. For server side computation, the Cortex-M0 implementation needs about 1.5 million cycles, and for the client side, 1.8 million. The corresponding cycles on the more potent M4 platform are 0.8 million and 9.8 million.

CHALLENGES AND FUTURE RESEARCH DIRECTIONS

Lattice-based encryption is usable even for devices with limited resources, according to a number of implementation findings. Lattice-based encryption is already more quickly computationally than conventional public-key cryptography, including RSA and even ECC. However, as lattice-based cryptography often involves greater communication costs, which uses considerably more resources, one cannot conclude that the former works better in reality. Lattice-based cryptography still has room for development. Of course, practical optimization is required, but theoretical progress toward shrinking the size of the ciphertext and signature may be more hopeful. Additional recommendations include more stringent security checks, efficient building methods, a decrease in the usage of discrete Gaussian noise, and effective encoding strategies.

Most of the systems we spoke about don't provide defence against side-channel assaults (SCAs). Since smart IoT applications are more vulnerable to side-channel attacks (SCAs), providing side-channel attack-resistant solutions is of utmost importance.

Most lattice-based encryption has a proven level of security, but this does not ensure actual security and may even make actual security less of a priority. Another difficulty is selecting acceptable settings for lattice-based methods. There are several, incomparable methods for studying lattice issues, and it is unclear how well some of them work. It would be ideal to have a single model for comparing the security of lattice-based encryption. The need for NTT-friendly options places additional restrictions on the parameters in R-LWE-based structures, which might result in security level gaps.

CONCLUSION

The various IOT and LWCRYPT approaches are employed since they are fairly lengthy and should be the subject of a separate investigation for better comprehension. Due to its effectiveness, LWCRYPT performs better for IOT security than other techniques. The use of less resource-intensive devices to connect to networks is made possible by cryptography. Concern for security will always exist anytime we develop new technology, including IOT. Even though there is nothing particularly novel about security, the methods of protection are altered by changes in platform and connectivity, power consumption, resource usage, and the force of trust placed in connected objects with minimal authorization and authentication checks. It is more than just a theoretical curiosity to examine the security of lattice-based encryption against complete quantum assaults. Currently, PQC solely takes the effects of quantum computers into account (or quantum algorithms). In the quantum realm, the attacker may interact quantumly with the cryptosystem, however, one kind of this approach is the quantum random oracle model seen in the literature. The most well-known and reliable block cypher is AES. AES is continuously being worked on to make it more portable and IoT-friendly. We want to continue working on the AES architecture in order to provide a lightweight IoT solution.

REFERENCES

[1]   B. Leukert, "IoT 2020: Smart and secure IoT platform", *International Electrotechnical Commission - White Paper*. Available From: http://www.iec.ch/whitepaper/iotplatform/ (Accessed on Nov 30th 2017).

[2]   S. Bruce, "IoT Cybersecurity: What's Plan B?", *Schneier on Security*. Available From: https://www.schneier.com/blog/archives/2017/10/iot cybersecuri.html (Accessed on Dec 12nd 2017).

[3]   E. Fernandes, A. Rahmati, K. Eykholt, and A. Prakash, "Internet of things security research: A rehash of old ideas or new intellectual challenges?", *IEEE Secur. Priv.*, vol. 15, no. 4, pp. 79-84, 2017. [http://dx.doi.org/10.1109/MSP.2017.3151346]

[4]   C. Cheng, R. Lu, A. Petzoldt, and T. Takagi, "Securing the internet of things in a quantum world", *IEEE Commun. Mag.*, vol. 55, no. 2, pp. 116-120, 2017. [http://dx.doi.org/10.1109/MCOM.2017.1600522CM]

[5]   Z. Liu, "High-performance ideal lattice-based cryptography on 8-Bit AVR microcontrollers", *ACM Transactions on Embedded Computing Systems3,* vol. 16, no. 4, 2017. [http://dx.doi.org/10.1145/3092951]

[6]   C. Peikert, "A decade of lattice cryptography", *Foundations and Trends® in Theoretical Computer Science,* vol. 10, no. 4, pp. 283-424, 2016. [http://dx.doi.org/10.1561/0400000074]

[7]   S. Roy, "Compact ring-LWE cryptoprocessor", *International Workshop on Cryptographic Hardware and Embedded Systems (CHES 2014)* Springer, Berlin, Heidelberg pp.371-391. year.2014.

[8]   J. Howe, "On practical discrete Gaussian samplers for lattice-based cryptography", *IEEE Trans. Comput.* 21 December 2016, pp. 322-334.

[9]   O.M. Guillen, "Towards post-quantum security for IoT endpoints with NTRU", *Design, automation & test in europe conference & exhibition* 27-31 March 2017, Lausanne, Switzerland.

[10] J. Buchmann, "High-performance and lightweight lattice-based public-key encryption", *In Proceedings of the 2nd ACM International Workshop on IoT Privacy, Trust, and Security,* pp. 2-9, 2016. [http://dx.doi.org/10.1145/2899007.2899011]

[11] T.G. Uneysu, and T. Oder, "Towards lightweight identity-based encryption for the post-quantu--secure internet of things", *In 18th International Symposium on Quality Electronic Design* 14-15 March 2017,Santa Clara, CA, USA pp. 319-324.

[12] P. oppelmann, and T. Guneysu, "Area optimization of lightweight lattice-based encryption on reconfigurable hardware", *In 2014 IEEE International Symposium on Circuits and Systems* 01-05 June 2014, Melbourne, VIC, Australia, pp. 2796-2799.

[13] J. Howe, T. Pöppelmann, M. O'neill, E. O'sullivan, and T. Güneysu, "Practical lattice-based digital signature schemes", *ACM Trans. Embed. Comput. Syst.,* vol. 14, no. 3, pp. 1-24, 2015. [http://dx.doi.org/10.1145/2724713]

[14] T.P. oppelmann, "Enhanced lattice-based signatures on reconfigurable hardware", In: *In International Workshop on Cryptographic Hardware and Embedded Systems* Springer: Berlin, Heidelberg, 2014, pp. 353-370.

[15] E. Alkim, "NewHope on ARM Cortex-M", In: *International Conference on Security, Privacy, and Applied Cryptography Engineering* Springer International Publishing, 2016, pp. 332-349. [http://dx.doi.org/10.1007/978-3-319-49445-6_19]

CHAPTER 15

Machine Learning For Traffic Sign Recognition

U. Lathamaheswari[1,*] and **J. Jebathagam**[1,2]

[1] *Department of Computer Science, Vels Institute of Science Technology and Advanced Studies, (VISTAS), Chennai, India*

[2] *Department of Information Technology, Vels Institute of Science Technology and Advanced Studies, VISTAS, Chennai, India*

Abstract: The recently developed technology in autos makes traffic signal prediction devices obligatory. It teaches users how to drive safely and manoeuvre their vehicles effectively. Due to drivers' various forms of attention, the number of accidents is rising alarmingly nowadays. The danger of distracted driving, which causes accidents, is decreased thanks to this technology, which also assists in identifying and providing information based on data. The notion of machine learning is presented, and the concepts of supervised learning, unsupervised learning, and reinforcement learning are covered under the heading of categorization and serve as the main principle. Linear regression, neural networks, naive Bayes, random forests, support vector machines, clustering, *etc.* are some types of models that machine learning may give. This study describes how to train a model using machine learning, with the basic principle being to divide the data into training, testing, and validation. The last section of this chapter discusses how to access machine learning methods to improve the quality of a machine learning project. The suggested approach provides an explanation of the combined model of the modern convolutional neural network (CNN) and the classic support vector machine (SVM) for traffic sign identification. Essentially, a CNN model was trained to produce this model. Several CNN model designs, including LeNet, AlexNet, and ResNet-50, may be used here. The subsequent layers of CNN's output may be utilised as features. These characteristics were added to SVM for categorization purposes.

Keywords: Convolutional neural network, Classification of signal image, Machine learning, Support vector machines.

INTRODUCTION

The electronic system of the car can distinguish a range of traffic symbols and signs on the road thanks to the classification and identification method used for

---

* **Corresponding author U. Lathamaheswari:** Department of Computer Science, Vels Institute of Science Technology and Advanced Studies, (VISTAS), Chennai, India; E-mail: lathasukumar1812@gmail.com

S. Kannadhasan, R. Nagarajan, N. Shanmugasundaram, Jyotir Moy Chatterjee & P. Ashok (Eds.)

traffic signals. It upholds traffic laws, ensures safety, and reduces the frequency of traffic collisions. The recommended method is advantageous for many real-world applications, such as driverless vehicles, driver assistance systems, and navigation systems.

Machine learning, a component of artificial intelligence, allows us to automatically examine data and build analytical models. It is based on the idea that computers can analyse data, spot trends, and reach conclusions either automatically or with little human intervention. The machine-analyzed data may be used as the main basis for making decisions.

Machine learning is used to optimise various methods for creating mathematical models and making decisions using knowledge or data from the past. It is presently used for a variety of tasks, such as speech recognition, recommender systems, Facebook auto-tagging, face recognition and image identification, email searching and filtering, and more [1 - 5].

How successfully a machine learning algorithm predicts a model depends on the quantity of data provided by the system and a large amount of data allows the machine to understand more information and improve its prediction accuracy. Fig. (**1**) depicts the data flow in the machine learning model.

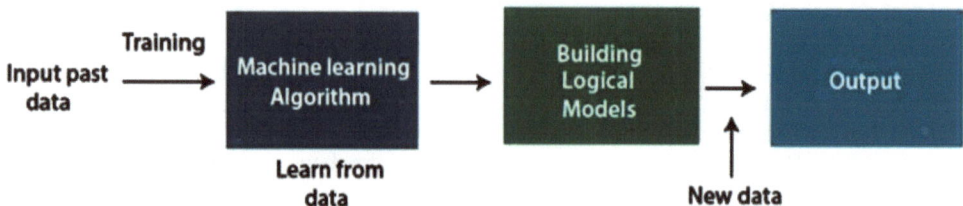

Fig. (1). Flow of data in machine learning model.

Loan Cristian Schuszter, "A Comparative Study of Machine Learning Methods for Traffic Sign Recognition, 2017.

WHY MACHINE LEARNING IS NEEDED?

The need for machine learning is increasing quickly as technology advances in line with user demands since this technology can automatically learn data without human intervention and can interpret complicated programmes that are challenging for humans to comprehend. We needed a computer to quickly correct the programmes using the data at hand since humans are incapable of manually optimising complicated data [6 - 10].

We may easily train the model using the machine learning algorithm by giving it a large amount of data, letting it handle the data, analysing the programme to use it to construct a particular model, and lastly automatically predicting the desired output.

Currently, it is utilised for a variety of purposes, including identifying online fraud, recommending Facebook friends, self-driving automobiles, virtual personal assistants, automatic language translation, email spam filtering, stock market exchange, *etc.*

CLASSIFICATION OF MACHINE LEARNING

Three major categories may be used to classify machine learning: supervised learning, unsupervised learning, and hybrid learning.

Supervised Instruction

It is a machine learning technique in which the system is trained using a collection of sample, labelled data to produce the desired outcome. The objective is to teach the computer by providing sample data, and then process the model to anticipate the outcome. It is always processed in a controlled environment. Spam filtering is one example of supervised learning.

Unsupervised Learning

Unsupervised learning is a technique that involves little to no human intervention while the computer learns.

The highlighted algorithm will automatically sort and take action on the data without any human intervention. The objective is to give new characteristics and identify patterns in the complicated data structure.

Some of the techniques used in unsupervised learning include principal component analysis, hierarchical clustering, and K means clustering.

Reinforcement Learning

A learning programme based on input from the environment is known as reinforcement learning. The agent will analyse the surroundings in this case and adjust the programme as necessary. With more favourable comments, the reward points will grow, and the performance will be enhanced as a result.

Machine learning models are often categorised into two categories: supervised learning and unsupervised learning, as illustrated in Fig. (**2**). The categorization of machine learning may be presented using a variety of techniques.

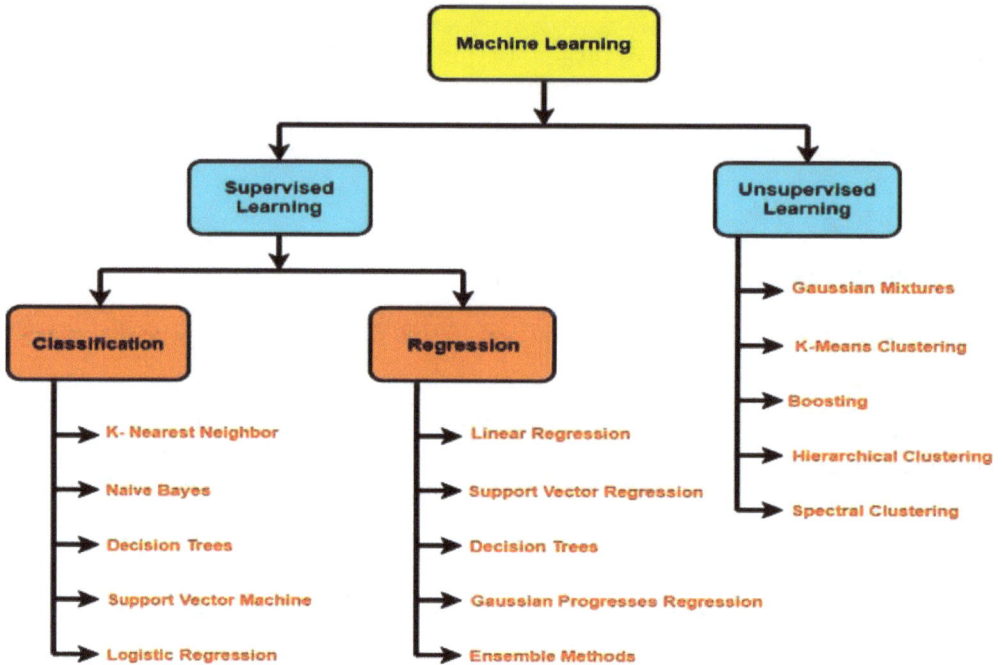

Fig. (2). Classification of machine learning model.

G. Meena, *et al.*, "Traffic Prediction for Intelligent Transportation System using Machine Learning", 2020.

MACHINE LEARNING IN TRAFFIC SIGN DETECTION

As a part of AI, machine learning aims to provide the model with minimal human interventions. Machine learning algorithms are applied in many applications to make use of AI for an accurate rate of desired results. It is achieved in many sectors [11 - 14]:

• Health

• Education

• Automobile

• Finance

• Education

• Gaming *etc*.

Common Factors of Road Accidents

Road accidents are common nowadays as it rapidly increases day by day. Fig. (**3**) shows a collection of road accident data.

Fig. (3). Road accidents due to remiss.

V.K. Kukkala, *et al.*, Advanced driver-asiistance systems: A path toward autonomous vehicles, 2018.

There are various reasons for the road accidents to occur which lead to major injuries causingdeath. Fig. (**4**) shows the rapid increase in road accidents day by day:

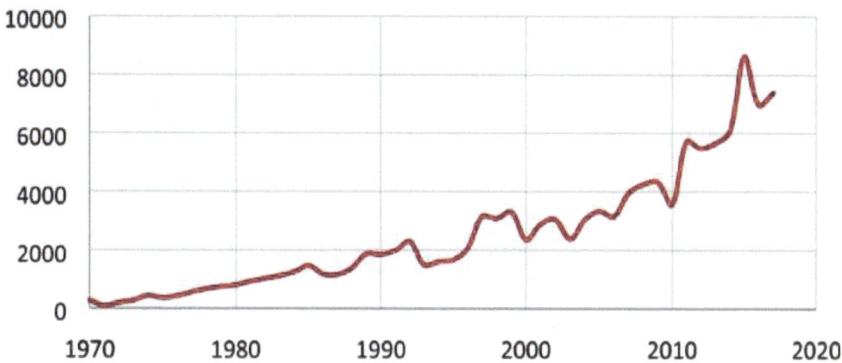

Fig. (4). Data of road accidents.

United Nations Road Safety Collaboration, Global plan for the decade of action for road safety, 2016.

x: number of years

y: number of persons involved in accidents

According to the facts, accident cases are rising substantially and alarmingly, thus it is critical to identify a precise strategy to lessen the causes.

Several contributing variables to traffic accidents include:

• Wrong turning; unsafe speed; auto accident; alcohol and drug use; incorrect interpretation of traffic signs and symbols.

The aforementioned elements may make road accidents inevitable since they depend on a person's prompt response, but technology might help to prevent them. Governments have implemented several steps to lower the risk of fatalities while driving, but with the aid of technology, the intended outcome will be more easily achieved.

United Nations Road Safety Collaboration, Global plan for the decade of action for road safety, 2016.

Fig. (**5**) shows the rise in accidents, which is mostly due to drivers' misconceptions and lack of attention. The most modern technologies are quite useful in directing drivers when they are driving on roadways in order to prevent such situations.

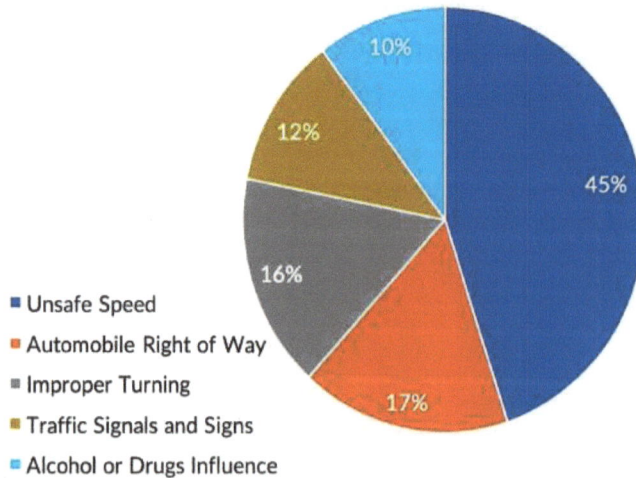

Fig. (5). Major causes of road accidents.

We can create a model for the detection and identification of traffic signs and symbols using AI and machine learning methods.

The old approaches, which arc focused on the colour and form of the objects, are insufficient for distinguishing actual signs from fraudulent ones when it comes to traffic sign identification and recognition in complex driving environments. These days, object identification and recognition are only a few of the computer vision applications that benefit from using machine learning techniques. Numerous experimental findings demonstrate that the fusion of the conventional SVM and deep neural CNN algorithms yields great levels of accuracy. This combination makes it simple to detect and recognise complex traffic signals. Fig. (6) shows the amount of traffic collected.

Fig. (6). Types of trafficsigns and symbols.

Syed Aley Fathima, *et al.*, "Object Recognition and Detection in RemoteSensing Images: A Comparative Study", 2020.

PROPOSED SYSTEM

The proposed system consists of two stages:

i. Detection stage

ii. Recognition stage

Detection Stage

In the detection stage, the image of traffic signals can be identified using CNN networks and the features are extracted for training and testing the data.

V.K. Kukkala, *et al.*, Advanced driver-assistance systems: A path toward autonomous vehicles, 2018.

Recognition Stage

The training and tested data are gathered and sent to a standard SVM classifier in the recognition step so it can categorise the features and provide the appropriate output. Fig. (7) illustrates how the data were used for training and testing. The author of a study [13] explicitly demonstrates the data that will be evaluated and trained in a machine learning model. Based on that, our proposed model can forecast and assign data for training and testing with the appropriate flow chart of the model, which is displayed in Fig. (8) and clearly describes the flow of data in our model.

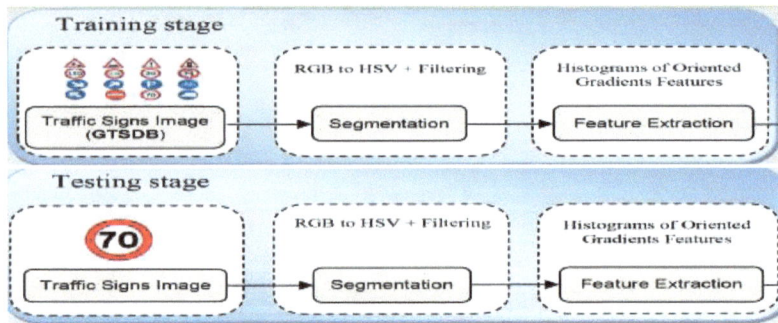

Fig. (7). Training and testing stage.

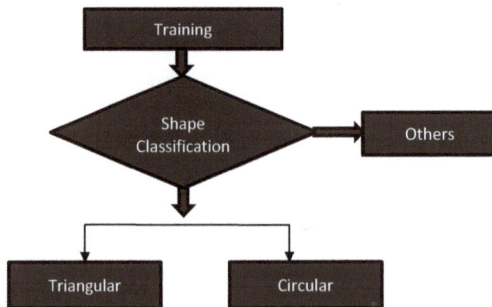

Fig. (8). Training and testing model.

V.K. Kukkala, *et al.*, Advanced driver-assistance systems: A path toward autonomous vehicles, 2018.

CONCLUDING REMARKS

The most extensive aspect of learning is machine learning, which teaches computers to execute tasks automatically. You must utilise metrics to gauge your model's success in order to assess your machine learning project. The metrics you employ will depend on the nature of the issue. Metrics like accuracy, precision, and others may be used to assess a classification model's performance. To get the required results, this data may be processed using a variety of machine-learning approaches. Machine learning is being used to advance many new technologies, and its use will undoubtedly increase in the future.

REFERENCES

[1]   C.S. Loan, "A comparative study of machine learning methods for traffic sign recognition", *19th International Symposium on Symbolic and NumericAlgorithms for Scientific Computing,* 21-24 September, Timisoara, Romania, 2017, pp. 389-392.

[2]   K. Gopala, P. Mohan, and S.D. Harihara, "Improved face recognition rate usinghogfeatures and svmclasifier", *J.Elect. Commun. Eng.,* pp. 886-893, 2016.

[3]   Z. Lin, H.K. Masafumi, Takigawa, and T. Kazuhiko, "Vehicle and pedestrian recognition usingmultilayer lidar based on support vector machine", *25th International Conference on Mechatronics and Machine Vision in Practice (M2VIP),* 20-22 November, Stuttgart, Germany, 2018, pp. 1-6.

[4]   C.C. Chang, and C.J. Lin, "LIBSVM: A library for support vector machines", *ACM Transactions on Intelligent Systems and Technology,* vol. 2, no. 3, pp. 1-27, 2011. Available From: http://www.csie.ntu.edu.tw/
[http://dx.doi.org/10.1145/1961189.1961199]

[5]   G. Meena, D. Sharma, and M. Mahrishi, "Traffic prediction for intelligent transportation system using machine learning", *In 3rd International Conference on Emerging Technologies in Computer Engineering: Machine Learning and Internet of Things,* 07-08 February, Jaipur, India, 2020, pp. 145-148.

[6]   A.F. Syed, A.K. Kumar, P. Ajay, and S.R. Syed, "Object recognition and detection in remotesensing images: A comparative study", *in International Conference on Artificial Intelligence and Signal Processing,* 10-12 January, Amaravati, India, 2020, pp. 1-5.

[7]   R. Girshick, "Fast R-CNN", *IEEE International Conference on Computer Vision (ICCV),* 07-13 December, Santiago, Chile, 2015, pp. 1440-1448.
[http://dx.doi.org/10.1109/ICCV.2015.169]

[8]   Li. Fei-Fei, and P. Peron, "A Bayesianhierarchical model for learningnaturalscenecategories", *IEEE Conference on Computer Vision and Pattern Recognition (CVPR 2005),* 20-25 June 2005, San Diego, CA, USA. pp.524-531.

[9]   B. Alexe, T. Deselaers, and V. Ferrari, "Measuring the objectness of image windows", *IEEE Trans. Pattern Anal. Mach. Intell.,* vol. 34, no. 11, pp. 2189-2202, 2012.
[http://dx.doi.org/10.1109/TPAMI.2012.28] [PMID: 22248633]

[10]   G. Cheng, and J. Han, "A survey on object detection in optical remote sensing images", *ISPRS J. Photogramm. Remote Sens.,* vol. 117, pp. 11-28, 2016.
[http://dx.doi.org/10.1016/j.isprsjprs.2016.03.014]

[11]   S.D Harihara, and P.G. Krishna Mohan, "Enhancement of face recognition rate by data base pre-processing", *Int. J.Comp. Sci. Inform. Technol.,* vol. 6, no. 3, pp. 2978-2984, 2015.

[12]　"United nations road safetycolaboration", *Global plan for the decade of action for road safety,* pp. 2011-2015, 2016.

[13]　V.K. Kukkala, J. Tunnell, S. Pasricha, and T. Bradley, "Advanced driver-asiistancesystems: A pathtowardautonomousvehicles", *IEEE Consum. Electron. Mag.,* vol. 7, no. 5, pp. 18-25, 2018. [http://dx.doi.org/10.1109/MCE.2018.2828440]

[14]　Y. Saadna, and A. Behloul, "An overview of traffic sign detection and classification methods", *Int. J. Multimed. Inf. Retr.,* vol. 6, no. 3, pp. 193-210, 2017. [http://dx.doi.org/10.1007/s13735-017-0129-8]

CHAPTER 16

Analysis of Machine Learning Algorithms in Healthcare

M. Nisha[1,*] and **J. Jebathagam**[1,2]

[1] *Department of Computer Science, Vels Institute of Science Technology and Advanced Studies, VISTAS, Chennai, India*

[2] *Department of Information Technology, Vels Institute of Science Technology and Advanced Studies, VISTAS, Chennai, India*

Abstract: Machine learning entails making changes to the systems that carry out artificial intelligence (AI)-related tasks. It displays the many ML kinds and applications. It also explains the fundamental ideas behind feature selection methods and how they can be applied to a variety of machine learning (ML) techniques, including artificial neural networks (ANN), Naive Bayes classifiers (probabilistic classifiers), support vector machines (SVM), K Nearest Neighbour (KNN), and decision trees, also known as the greedy algorithm.

Keywords: Algorithms, Artifical intelligence, Classification, Learning methods, Machine learning, Svm.

INTRODUCTION

Machine learning is the study of algorithms by computers that can only become better with time or with the use of data. Machine learning heavily includes AI [1 - 5]. Machine learning algorithms learn knowledge directly from the provided data rather than using a model that has been preset. The algorithm categorization is shown in Fig. (1).

CLASSIFICATION OF MACHINE LEARNING

Unsupervised Learning

It's a kind of algorithm that discovers patterns from data that hasn't been labelled, or that hasn't been tagged. As a result, the machine is used to develop its own internal representation of its environment and to produce the crucial material from

* **Corresponding author M. Nisha:** Department of Computer Science, Vels Institute of Science Technology and Advanced Studies, VISTAS, Chennai, India; E-mail: manikantnisha23@gmail.com

S. Kannadhasan, R. Nagarajan, N. Shanmugasundaram, Jyotir Moy Chatterjee & P. Ashok (Eds.)

it. This is a crucial process that is copied from learning with humans. In unsupervised machine learning, clustering the data into patterns does not provide any outputs that can be predicted. Its performance is assessed after completing supervised machine learning exercises. Fig. (**2**) displays the flow diagram for machine learning in the cardiovascular field. Unsupervised machine learning leads to a state called heterogeneous, which is analogous to an abrupt myocardial infarction arrest. Once again, we may begin with a big collection of numbers that are representative of every person with unexplained heart failure [6 - 10]. In order to characterise the composition of cells' complex biomolecules in each sample, one may conduct Myo-Cardic biopsies and use several immunological staining techniques. In contrast to supervised learning, there is no expected result. Thus, we may demonstrate the data's interest in identifying patterns. In reality, this presents an SL difficulty since, when a model is developed and patients are categorised according to risk, there is a potential that the identification of the illness may be lost.

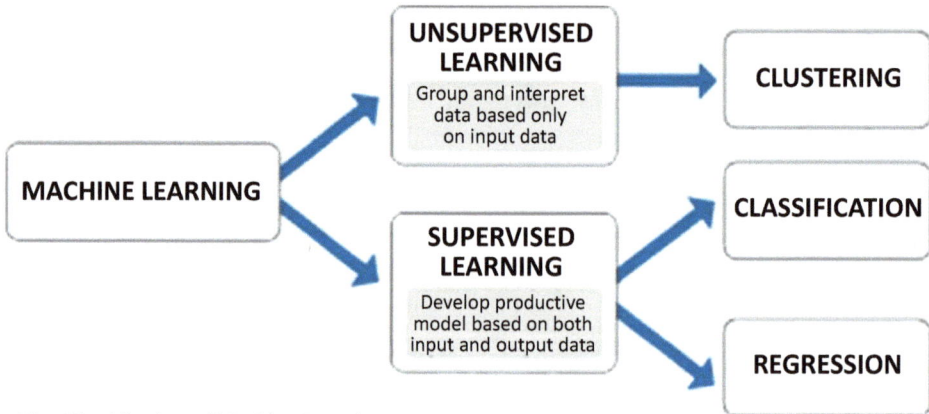

Fig. (1). Classification of Machine Learning.

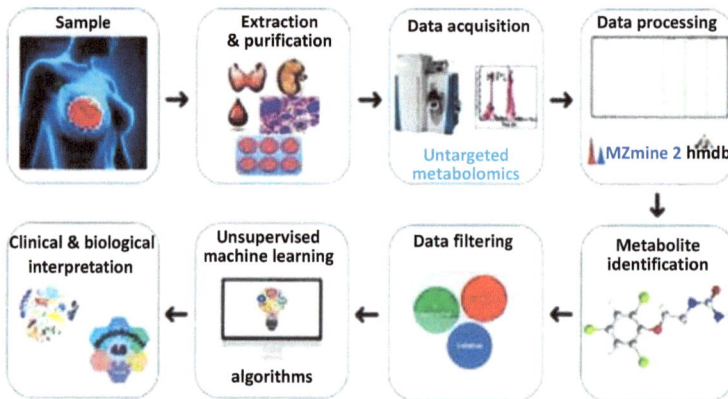

Fig. (2). Machine Learning in Cardiovascular.

Supervised Learning

A subset of both machine learning and artificial intelligence is supervised learning. It is used to train algorithms or to properly anticipate its data by categorising the labelled data sets. Its first objective is to forecast the algorithm's known aim and output [11 - 15]. There are many higher competitions in machine learning, so each participant is evaluated on their own merits using data sets from repeated supervised learning tasks that include handwriting recognition, object classification using images, and document classification. Fig. (**3**) depicts the labelled and unlabeled training algorithm. SL concentrates on its own categorization, which requires it to choose a subset for itself and to select the best new instance of data, which it uses to forecast. It also requires to estimate an unidentified parameter. In order for a trained individual to function successfully, it is often tried to establish a steady performance on humans. An ECG (Electrocardiogram) signal is used to gauge the heart's activity. The use of ML-libs is being made to put the suggested strategy into practise. The scalable machine learning library for Apache Spark is called ML-lib. Its characteristics serve as the ML algorithm's inputs. The automatic pattern recognition that may be chosen to execute a restricted selection of diagnoses is how the ECG is interpreted. In supervised machine learning, automated detection is used to identify a nodule from a chest x-ray that would also serve as a radiography [16 - 20].

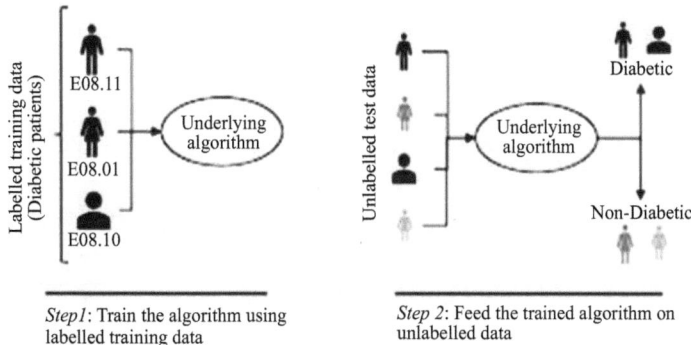

Step1: Train the algorithm using labelled training data   *Step 2*: Feed the trained algorithm on unlabelled data

Fig. (3). ECG in machine learning.

Machine Learning in Health Care

It often includes an ML and SL technique that is used to train the data with labels for the training models in order to forecast an illness that is a danger. The test is administered in sets of high and low-risk patients. These models, however, are exclusively employed for research and clinical applications [21 - 25]. Fig. (**4**) illustrates how machine learning algorithms are employed in the healthcare industry.

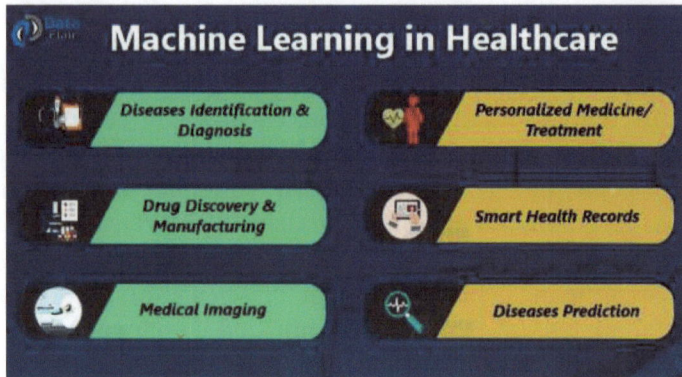

Fig. (4). ML algorithm in health care.

Drug Discovery in ML

Companies like Pharmaceutical have profited immensely from using ML algorithms in their drug development processes [26]. In order to identify and forecast compounds that are employed in drug development, ML algorithms have been used to construct a variety of types that predict chemical, physical, as well as biological aspects of compounds. Fig. (**5**) depicts the pharma industry's step-b--step procedure. It is possible to include the drug discovery process. Therefore, ML algorithms have been used to uncover novel medication uses that can also predict interactions between pharmaceuticals and proteins, therapeutic effectiveness, and the safety of biomarkers, among other things.

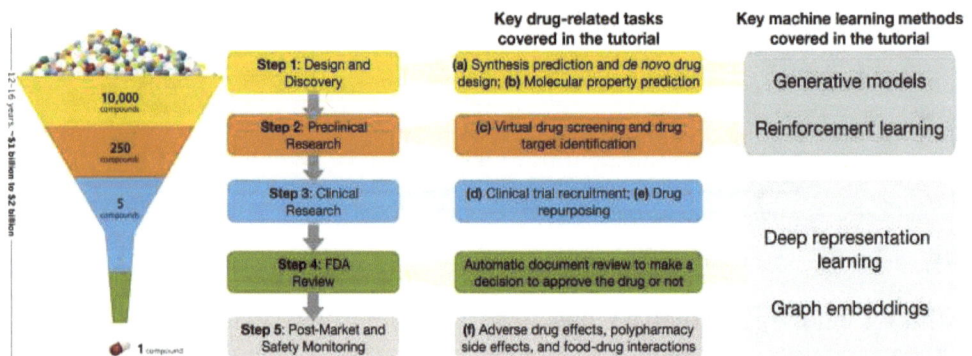

Fig. (5). Process in drug industry.

ML in Medical Imaging

Medical pictures are found using ML techniques, which may also be utilised to spot trends. Fig. (**6**) identifies machine learning in image processing for us. Even though it is a great tool for making medical diagnoses, it may nevertheless be misused. Its technique often starts with ML system computing with picture characteristics that might be thought of as being important in generating a diagnostic or making a forecast [27]. The ML algorithm system is used to find the optimal combination of the image characteristics for categorising images or calculating certain attributes as needed for the given particular area of the picture. Each may be utilised in a variety of ways, with the aid of various machine-learning techniques that make testing simpler and aid in applying the photos. Future generations will be more impacted by the usage of medical imaging in machine learning algorithms. Anyone interested in working in medical imaging should be aware of the benefits, drawbacks, and operation of machine learning [26 - 30].

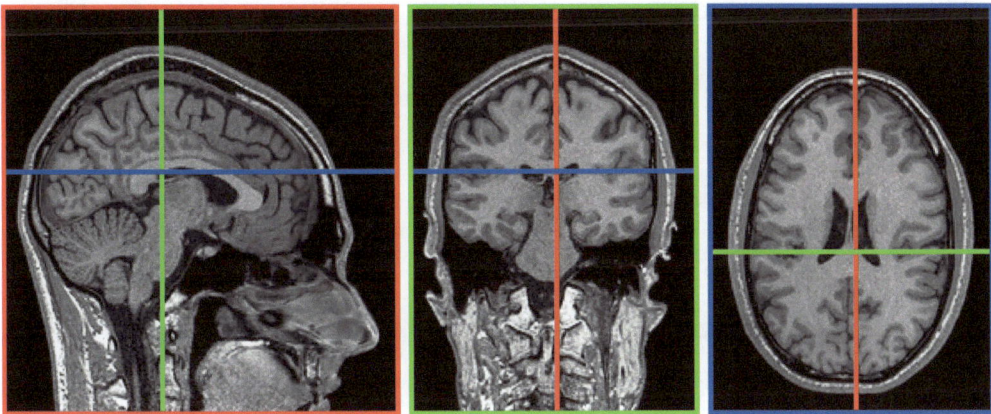

Fig. (6). Identifying Imaging process in Machine learning.

ML in Smart Health Records

In this real-world situation, machine learning is a high-level technique for implementing any form of healthcare. To find the best predictions for the patients in their specific health conditions, we need to use machine learning approaches, and we also need to examine past medical information. For that, we must have a repository or warehouse where we can save digital information on the patients and their care. In this article, we propose a programme for maintaining patient health data utilising a digital card that can only be accessed by a doctor, a receptionist, and the respected staff members of other hospital departments. The information that will be gathered from patients is shown in Table **1**. We used a responsive web application to execute this prediction approach, and we used machine learning and Python to build the other components, such as forecasts and statistical analysis. In this article, we use certain machine learning approaches to our data in order to identify the best course of action for the therapeutic process and the proper upkeep of the patient's digital records. The record we save will be in chronological sequence, and they may be changed in the future with admin access. All of this data was kept in the form of an application, and we described how to use machine learning in this application as well as how to create and use prediction models. ML will be the primary strategy for any healthcare project that can be carried out in the actual world. For so, we must investigate ML approaches in order to pinpoint the best prediction values that may be linked to patients' current health conditions, as well as scrutinise their prior medical histories [28]. To do this, a record must be kept in which we must track all digital information pertaining to patient therapy. Fig. (7) depicts the many applications used to keep patient health information utilising digital cards that can only be used by the hospital's doctors, receptionists, and other recognised staff members. Using ML algorithms, certain techniques will determine the data and the best solution, conduct the therapy, and preserve the health of the patient's digital records. The administrator will keep these records in chronological order, and there is an opportunity for data modification.

Table 1. Machine Learning for Smart Health Records.

Age	Reports
Weight	ScannedReports
Height	History of medication
Contact Info	PresentMedication
Habits	PresentScanreports(IfAny)
Previous Disease Information	PresentLabReports(IfAny)
Current Symptoms	Present update
Duration of the suffering	Next update

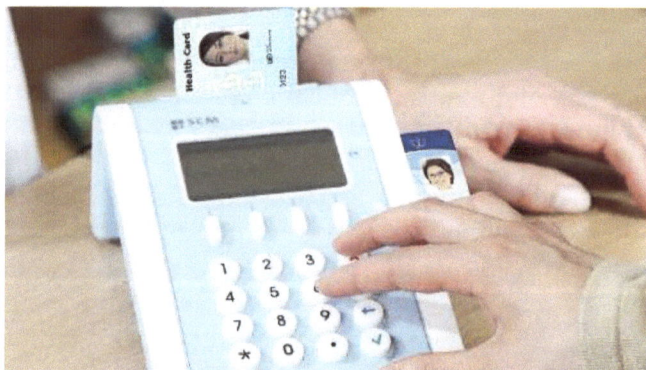

Fig. (7). Smart digital cards in machine learning.

Computer Science and ML

ML produced tremendous results in computer science. Building computers that could solve specific problems has been the emphasis of computer science, which also aims to determine if a problem can be solved or not. When fresh data is introduced to a computer that has been manually programmed, the difficulty is that the machine will automatically re-program itself based on its original learning.

Learning From Both Humans and Machines

A computer may learn from a human or animal mind in terms of experience and time utilising learning techniques. Machine learning is utilised to address issues with the human brain. The difference between animal and human psychology is still a mystery. Machine learning is used to examine various learning approaches because of the differences between machines and humans [31 - 37].

EXAMINING DATA, MLAND AI

Data mining, which encompasses both AI and ML, is explains pretty much any kind of data. The mined data may be used as the foundation for both ML and AI, thus it is not only a little approach for proving hypotheses, but also a way to developing hypotheses that are meaningful. Therefore, it is possible to describe artificial intelligence as having the capacity to resolve a particular issue without the assistance of a person. The algorithm employed in data mining is what drives the interpretation process. There are instances when the system provides solutions that are not expressly planned, but it is important for the data to provide a solution on its own. By supplying the data to train on and change in order to expose the specific data to new data, machine learning is brought to a higher level. In order to increase performance and create new data with the greatest outcomes, it focuses on extracting information from the many data sets that are used to discover and identify patterns. It is quite improbable to create any form of computer with intelligence, such as language or vision.

SL Use of Unlabeled Data

Approaching the fact that the data won't always be accessible is the biggest issue. First, utilising unsupervised learning, the data needs to be pre-processed, labelled, and filtered. To do this, we must apply supervised classification for feature extraction, dimension reduction, *etc.* Therefore, supervised learning is used to categorise web pages or find spam emails. Current learning challenges or new algorithms make use of the effectiveness of unlabeled data.

Linking Different Ml Algorithms

Numerous machine learning algorithms have been introduced and tested in a wide range of areas [30]. The examples and scenarios of a certain Ml algorithm have been applied to two supervised classification methods: logistic regression and naive Bayes. These two have a large number of data sets, but only a few are utilised as training data for implementation.

Similarity-based Grouping of Algorithms

Regression Algorithms (RA) are a kind of predictive analytics that make use of the relationship between independent and dependent variables. Fig. (**8**) displays the many regression types, which include.

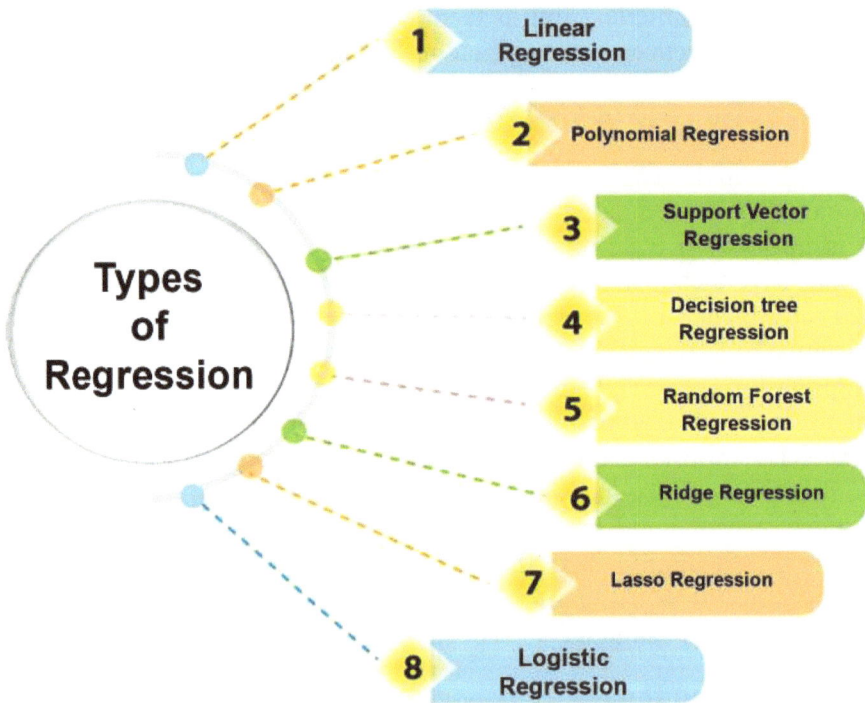

Fig. (8). Types of Regression.

Instance-based Algorithms

It is built on a learning model, as shown in Fig. (**9**), which stores training data for each occurrence rather than using them to construct a precise specification of the intended goal function. Every time a new issue is raised or produced, it may be used to investigate according to an example, such as [31], which can be saved in order to calculate or forecast the precise target function value. An existing example can also be easily changed by a new one that has been stored. Fig. (**9**) depicts the memory-based approach, which is another name for it.

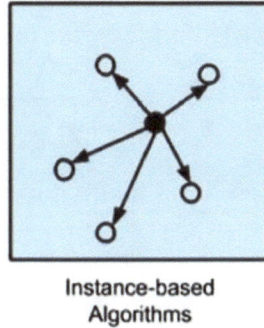

Instance-based
Algorithms

Fig. (9). Based on learning model.

Regularization Algorithm

It is a simple algorithm but a powerful modification, which is augmented with the other ML existing models typically Regressive Models [32]. It smooths up the regression line formally with any bend of the curve that tries to match the outliers of the models. The regularization of ML types is shown in Fig. (**10**).

Types of Regularization in ML

Fig. (10). Types of regularization.

Decision Tree Algorithms

It constructs a tree like structure which also involves a possible solution to a problem based on certain constraints, and it is named as a simple single decision or a root, until a decision or prediction is made to form a tree [33]. Fig. (**11**) shows the structure of the decison tree algorithm.

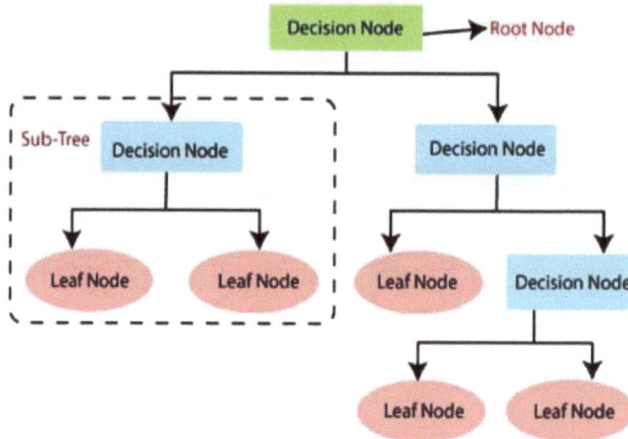

Fig. (11). Structure of decisión tree algorithm.

Bayesian Algorithms

A collection of Bayes Theorem-based Machine Learning (ML) algorithms is used to address classification and regression issues.

Support Vector Machine (SVM)

It is the most well-known method in machine learning that can stand alone. It is used to distinguish between a hyperplane, a decision plane, or a decision boundary among a group of data points to categorise with various labels [34]. The plots linked inside the SVM classification plane are shown in Graph 1. A supervised classification method serves as its foundation. SVM can be classified in both linear and nonlinear ways.

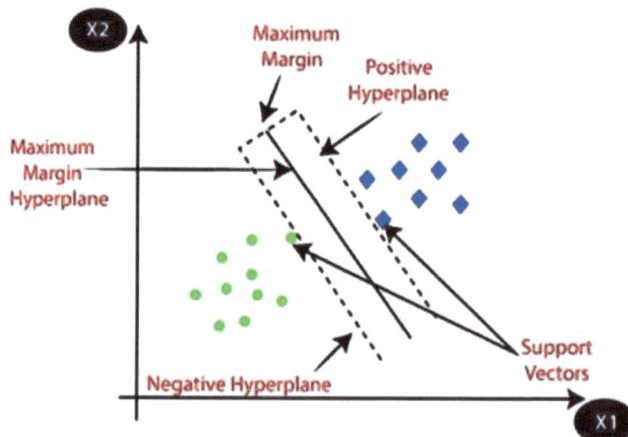

Graph 1. Based on SVM classification.

Clustering Algorithms

Using a firmly determined pattern, cluster datasets to categorise and label the relevant data.

Association Rule Learning Algorithms

The frequently used association rules that aid to identify correlation between the unassociated data [35] help to find a correlation between the unassociated data. They are often used in e-commerce websites to forecast the behaviour of each consumer and his future demands in order to market specific items to him.

Artificial Neural Network (ANN) Algorithms

The activities that create real neural networks in humans or other animals are known as non-linear modelling processes, and they serve as the foundation for this model. However, it looks for connections between the data intake and output. However, it uses sample data rather than taking into account the complete dataset to save time.

Ensemble Algorithms

These kinds of instructional methods for learners should be carefully selected to maximise accuracy.The ensemble of approaches is shown in Fig. (**12**). The basic goal of the ensemble technique is to combine the many estimates made by weaker estimators.

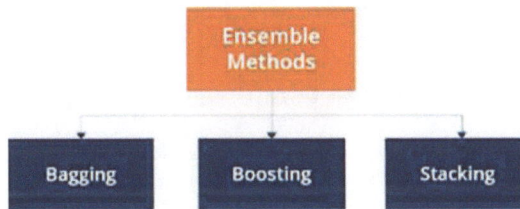

Fig. (12). Methods in Ensmeble Algorithm.

Applications

ML is an important real-life applications used in day to day life, some of which are explained here briefly.

Speech Recognition

All speech recognition systems use ML to approach the trained system for better accuracy [36]. Most of such systems learning has distinct phases which is shown in Fig. (**13**).

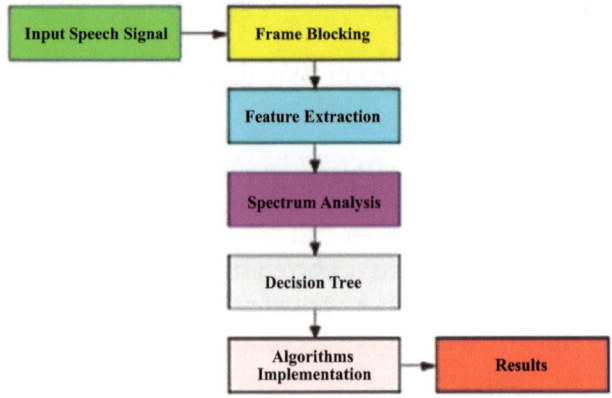

Fig. (13). Different phases in speech recognition.

Computer Vision

The majority of modern system views use machine learning (ML) to improve accuracy and can classify microscopic images of cells automatically. These systems are similar to face recognition software.

Bio-Survelliance

AnML software is taught to recognise common symptoms, patterns, and local distribution using the patient profiles of those who have been hospitalised. Several government programmes were launched to use ML algorithms to monitor disease outbreaks [37]. Only automated learning techniques can handle complicated and dynamic data sets like these. Fig. (**14**) illustrates the application of data analysis in biosurveillance.

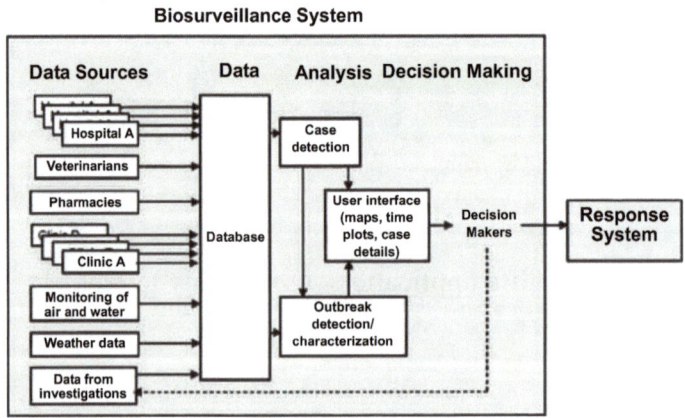

Fig. (14). Data Analysis of Biosurvelliance.

Control or Robot for Automation

Google's self-driving vehicles learn from acquired topographical data using machine learning. Robotic and automated systems use machine learning techniques the most.

Experiments in Empirical Science

ML is being used in astronomy, genetics, neuroscience, and psychological analysis to distinguish between common and exceptional celestial objects. Many coordinated intense data science projects apply ML techniques in a variety of their own research projects. The second small-scale ML application is significant and covers the following tasks: spam filtering, fraud detection, topic identification, and prediction analytics.

CONCLUDING REMARKS

The primary goal of ML research is to develop effective general-purpose learning techniques that are efficient and practical and can perform better in their respective fields. A data-driven algorithm has an overall total manual or direct programming and has the capacity to quickly review a huge quantity of data. Efficiency in using all available data resources is crucial for performing a paradigm, along with time and space complexity, and because they are frequently more accurate, and interpretable prediction rules are also of utmost importance, excluding human bias.

By creating new software to handle perception duties for sensors including proximity, temperature, computer vision, and voice recognition, anyone can assess or label a picture of a letter or by the alphabet with ease, but it is quite tough to carry out such work and to create an algorithm for this.

REFERENCES

[1]   T.M. Mitchell, *MachineLearning.* McGraw-HillInternational, 1997.

[2]   T.M. Mitchel, *The discipline of machine learning.,* Carnegie Mellon University, 2006.

[3]   N. CristianiniandJ, and T. Shawe, *Shawe-Taylor.An Introduction to Support Vector Machines.* Cambridge University Press, 2000.

[4]   E.F Osuna, and F. Girosi, "Support vectormachines: Training and applications", 1997.

[5]   V. Vapnik, *StatisticalLearningTheory.* JohnWiley&Sons, 1998.

[6]   C.J.C. Burges, "Atutorialon support vector machines for pattern recognition", *Data Mining and Knowledge Discovery,* vol. 2, no. 2, pp. 1-47, 1998.

[7]   A. Taiwo Oladipupo, "Types of machine learning algorithms", In: *New Advances in Machine Learning,* Z. Yagang, Ed., InTechopen, 2010.

[8]   T. Mitchell, W. Cohen, E. Hruschka, P. Talukdar, J. Betteridge, A. Carlson, B. Dalvi, M. Gardner, B.

Kisiel, J. Krishnamurthy, N. Lao, K. Mazaitis, T. Mohamed, N. Nakashole, and E. Platanios, "Proceedings of the Twenty-NinthAAAI Conferenceon Artificial Intelligence", *Never-Ending Learning,* 2014.

[9]   Pedregosa, "Scikit-learn", *Machine Learning in Python,* vol. JMLR12, pp. 2825-2830, 2011.

[10]   J. Wang, T. Jebara, and S. Chang, "F.Semi-supervisedlearningusinggreedymax-cut", *J. Mach. Learn. Res.,* vol. 14, no. 1, pp. 771-800, 2013.

[11]   O. Chapelle, V. Sindhwani, and S.S. Keerthi, "Optimization techniques for semi- supervised support vector machines", *J. Mach. Learn. Res.,* vol. 9, pp. 203-233, 2013.

[12]   J. Baxter, T. Jebara, and S.F. Chang, "Semi-supervisedlearningusinggreedymax-cut", *J. Mach. Learn. Res.,* vol. 12, pp. 149-198, 2000.

[13]   D. S.Ben, and R. Schuller, "Exploiting task relatedness for multiple task learning", *In Conference on Learning Theory,* 2003.

[14]   W. Dai, G. Xue, Q. Yang, and Y. Yu, "Transferring Naive Bayes classifiers fortextclassification", *AAAI Conference on Artificial Intelligence,* 22 July 2007,vol.1, pp. 540-545.

[15]   H. Hlynsson, *Transfer learning using the mini mum description length principle with adecision tree application.* Master'sthesis,University of Amsterdam, 2007.

[16]   Z. Marx, M. Rosenstein, L. Kaelbling, and T. Dietterich, *Transfer learning with anensemble of background tasks.* In NIPS Work shop on Transfer Learning, 2005.

[17]   R. Conway, and D. Strip, "Selective partial access to a data base", *In Proceedings of ACM Annual Conference,* pp. 85-89, 1976.

[18]   P.D. Stachour, and B.M.T. huraisingham, "Design of LDVA multi level secure relation al data base management system IEEE Trans", *Knowledge and Data Eng.,* vol. 2, no. 2, pp. 190-209, 1990.

[19]   R. Oppliger, "Internet security", *Commun. ACM,* vol. 40, no. 5, pp. 92-102, 1997. [http://dx.doi.org/10.1145/253769.253802]

[20]   A. Rakesh, and K.S. Rama, "Privacy preserving data mining", *SIGMOD'00 Proceedings of the 2000 ACMSIGMOD international conference on Management of data,* vol. 29, no. 2, pp. 439-450, 2000.

[21]   A. Carlson, J. Betteridge, B. Kisiel, B. Settles, E.R. HruschkaJr, and T.M. Mitchell, "Toward an architecture for never-ending language learning", *AAAI,* vol. 5, no. 3, 2010.

[22]   X. Chen, A. Shrivastava, and A. Gupta, "Neil:Extracting visual knowledge from web data", *In Proceedings ofICCV* 01-08 December, Sydney, NSW, Australia, 2013, pp. 1409-1416.

[23]   P. Donmez, G. Carbonell, and A. Gupta, "Proactive learning: Cost-sensitive active learningwith multiple imperfectoracles", In: *InProceedings of the 17thACM conference on In-formationand knowledge management,.* Napa Valley, California, in October 2008, pp.619-628.

[24]   T. M. Mitchell, J. Allen, P. Chalasani, J. Cheng, O. Etzioni, M.N. Ringuette, and J.C. Schlimmer, "Theo: Aframe work for self-improving systems", *Arch.forIntelli-gence,* pp. 323-356, 1991.

[25]   M.A. Ahmad, C. Eckert, and A. Teredesai, "Interpretable machine learning in healthcare", *Proceedings of the 2018 ACM international conference on bioinformatics, computational biology, and health informatics,* 04-07 June 2018,New York, NY, USA, pp. 559-560.

[26]   S. Dara, S. Dhamercherla, S.S. Jadav, C.H.M. Babu, and M.J. Ahsan, "Machine learning in drug discovery: A review", *Artif. Intell. Rev.,* vol. 55, no. 3, pp. 1947-1999, 2022. [http://dx.doi.org/10.1007/s10462-021-10058-4] [PMID: 34393317]

[27]   A. Sungheetha, and R. Sharma, "3D image processing using machine learning based input processing for man-machine interaction", *J. Innov. Imag. Proces.,* vol. 3, no. 1, pp. 1-6, 2021. [http://dx.doi.org/10.36548/jiip.2021.1.001]

[28]   A. Dhillon, and A. Singh, "Machine learning in healthcare data analysis: A survey", *J. Biol. Todays*

World, vol. 8, no. 6, pp. 1-10, 2019.

[29] A. Chanaa, "Deep learning for a smart e-learning system", *In 2018 4th International Conference on Cloud Computing Technologies and Applications,* 26-28 November 2018, Brussels, Belgium, pp. 1-8.
[http://dx.doi.org/10.1109/CloudTech.2018.8713335]

[30] N.J. Wahl, "An overview of regression testing", *Softw. Eng. Notes,* vol. 24, no. 1, pp. 69-73, 1999.
[http://dx.doi.org/10.1145/308769.308790]

[31] D.W. Aha, D. Kibler, and M.K. Albert, "Instance-based learning algorithms", *Mach. Learn.,* vol. 6, no. 1, pp. 37-66, 1991.
[http://dx.doi.org/10.1007/BF00153759]

[32] S. Okser, T. Pahikkala, A. Airola, T. Salakoski, S. Ripatti, and T. Aittokallio, "Regularized machine learning in the genetic prediction of complex traits", *PLoS Genet.,* vol. 10, no. 11, p. e1004754, 2014.
[http://dx.doi.org/10.1371/journal.pgen.1004754] [PMID: 25393026]

[33] B. Charbuty, and A. Abdulazeez, "Classification based on decision tree algorithm for machine learning", *J. Appl.Sci. Technol. Trend.,* vol. 2, no. 1, pp. 20-28, 2021.
[http://dx.doi.org/10.38094/jastt20165]

[34] S. Wang, X. Guo, Y. Tie, I. Lee, L. Qi, and L. Guan, "Graph-based safe support vector machine for multiple classes", *IEEE Access,* vol. 6, pp. 28097-28107, 2018.
[http://dx.doi.org/10.1109/ACCESS.2018.2839187]

[35] J. Bai, S. Xiang, and C. Pan, "A graph-based classification method for hyperspectral images", *IEEE Trans. Geosci. Remote Sens.,* vol. 51, no. 2, pp. 803-817, 2013.
[http://dx.doi.org/10.1109/TGRS.2012.2205002]

[36] L. Deng, and X. Li, "Machine learning paradigms for speech recognition: An overview", *IEEE Trans. Audio Speech Lang. Process.,* vol. 21, no. 5, pp. 1060-1089, 2013.
[http://dx.doi.org/10.1109/TASL.2013.2244083]

[37] S. Raouf, C. Weston, and N. Yucel, "Registered report: Senescence surveillance of pre-malignant hepatocytes limits liver cancer development", *eLife,* vol. 4, p. e04105, 2015.
[http://dx.doi.org/10.7554/eLife.04105] [PMID: 25621566]

SUBJECT INDEX

A

Acts, virtual machine monitor 88
Agricultural 113, 116
  guidelines 116
  manufacturing 113
Agricultural sensors 115, 116
  smart 116
Agriculture 113, 116
  activity 113
  sector 116
Aid 1, 2, 3, 118, 135, 139, 158, 187, 196, 202
  mobility 158
  visual 135
Air conditioner system 41
Algorithms 35, 90, 91, 93, 99, 159, 171, 172, 173
  digital image processing 159
  dynamic 93
  lattice-based 173
  learning 99
  load-balancing 91
  load-managing 90
  public-key encryption 171
  robust segmentation 35
  traditional 172
Angiography 148
Applications 4, 76, 126
  making cloud-based 76
  managing online business 126
  real-world automotive 4
Approaches, haematological 16
Artificial neural network (ANN) 15, 108, 192, 202
Assistive technology, innovative 158
Auto-correlation function 21
Automatic language translation 184
Autonomous system 15

B

Ballot paper 104, 106, 107
  method 106, 107
  system 104
  verification technique 104
Bayes theorem-based machine learning 201
Bayesian algorithms 201
Biometric 126, 132
  system 126
  technologies 132
Brain, human 109, 198

C

Camera, mobile 103
Cardiac plaque 156
Cardiovascular diseases 148
Cartridge-scanner system, quantitative 15
CCTA Image 149, 152, 153
Cloud 75, 76, 81, 85, 87, 88, 89, 90, 91, 92, 93, 94, 95, 96, 97
  community 88
  environment 76, 85, 87, 89, 93, 94, 96
  hybrid 88
  infrastructure 81
  private 88
  public 88
  service providers (CSP) 87, 90, 92, 95
Cloud computing 74, 76, 81, 87, 88, 89, 91, 93, 95, 96, 97
  architecture 88
  environment 81
  systems 81
  technology 76
Clustering 182, 184, 193, 202
  algorithms 202
  hierarchical 184
CNN networks 189
Commission, electoral 103, 105
Computer 5, 16, 109, 148, 150, 203, 204